Defense Planning and Readiness of North Korea

How has North Korea developed and managed its military readiness to achieve its strategic ends?

Hinata-Yamaguchi analyzes North Korea's defense planning by looking at how political, economic, and societal factors affect the Korean People's Army's (KPA) readiness and strategies. He answers four key questions: How have the internal and external factors shaped North Korea's security strategy? How do the political, economic, societal, and environmental factors impact North Korea's defense planning? What are North Korea's defense planning dilemmas, and how do they impact the KPA's readiness? What are the key implications for regional security and the strategies against North Korea?

This analysis, drawing on various Korean, English, Japanese, and Chinese sources on North Korea and military affairs, will be of great value to strategists and policy analysts as well as scholars of East Asian security issues.

Ryo Hinata-Yamaguchi is an adjunct fellow at the Pacific Forum, a visiting professor at the Pusan National University, and a project assistant professor at the Research Center for Advanced Science and Technology at the University of Tokyo.

Routledge Advances in Korean Studies

42 **Korean Adoptees and Transnational Adoption**
Embodiment and Emotion
Jessica Walton

43 **Digital Development in Korea, Second Edition**
Lessons for a Sustainable World
Myung Oh and James F. Larson

44 **The State, Class and Developmentalism in South Korea**
Development as Fetish
Hae-Yung Song

45 **Development Prospects for North Korea**
Edited by Tae Yong Jung and Sung Jin Kang

46 **The Road to Multiculturalism in South Korea**
Ideas, Discourse, and Institutional Change in a Homogenous Nation-State
Timothy Lim

47 **Healing Historical Trauma in South Korean Film and Literature**
Chungmoo Choi

48 **Exporting Urban Korea?**
Reconsidering the Korean Urban Development Experience
Edited by Se Hoon Park, Hyun Bang Shin and Hyun Soo Kang

49 **Defense Planning and Readiness of North Korea**
Armed to Rule
Ryo Hinata-Yamaguchi

For more information about this series, please visit: www.routledge.com/asian studies/series/SE0505

Defense Planning and Readiness of North Korea

Armed to Rule

Ryo Hinata-Yamaguchi

Routledge
Taylor & Francis Group
LONDON AND NEW YORK

First published 2021
by Routledge
2 Park Square, Milton Park, Abingdon, Oxon OX14 4RN

and by Routledge
52 Vanderbilt Avenue, New York, NY 10017

Routledge is an imprint of the Taylor & Francis Group, an informa business

British Library Cataloguing-in-Publication Data
A catalogue record for this book is available from the British Library

Library of Congress Cataloging-in-Publication Data
Names: Hinata-Yamaguchi, Ryo, author.
Title: Defense planning and readiness of North Korea : armed to rule / Ryo Hinata-Yamaguchi.
Description: Abingdon, Oxon ; New York : Routledge, [2021] | Series: Routledge advances in Korean studies | Includes bibliographical references and index.
Identifiers: LCCN 2020051591 (print) | LCCN 2020051592 (ebook) | ISBN 9780367482862 (hardback) | ISBN 9780367771102 (paperback) | ISBN 9781003039051 (ebook)
Subjects: LCSH: Military planning—Korea (North) | Korea (North)—Armed Forces—Operational readiness. | Korea (North)—Military policy. | Korea (North)—Defenses.
Classification: LCC U155.K7 H55 2021 (print) | LCC U155.K7 (ebook) | DDC 355/.03355193—dc23
LC record available at https://lccn.loc.gov/2020051591
LC ebook record available at https://lccn.loc.gov/2020051592

ISBN: 978-0-367-48286-2 (hbk)
ISBN: 978-0-367-77110-2 (pbk)
ISBN: 978-1-003-03905-1 (ebk)

Typeset in Galliard
by Apex CoVantage, LLC

Contents

Acknowledgments

Everything I have accomplished in my career, including this work, was made possible with the support from my family, friends, colleagues, mentors, and institutions who are too numerous to list here.

This effort was made possible with the constructive and insightful advice from my colleagues, friends, and mentors. In particular, I thank Carl W. Baker, Ralph A. Cossa, James Cotton, Paul Dibb, Brad Glosserman, Andrei Lankov, Bernard Fook Weng Loo, and Michishita Narushige for their guidance and mentorship over the years and for their comments that were critical to this study. Special thanks also to my great friend Collin Koh Swee Lean for all the chats, insights, and suggestions at various critical junctures. I am also deeply indebted to an unnamed Admiral for inspiring me to pursue my career in security and defense affairs, and for all his mentorship throughout the years.

Without the Pacific Forum, my career and this project would not have been possible. I particularly thank Carl W. Baker, Ralph A. Cossa, Robert P. Girrier, Brad Glosserman, James A. Kelly, and Lloyd R. "Joe" Vasey for their mentorship. I also especially thank Ellise Fujii Akazawa, Georgette Almeida, An Sun-na, Mari Ching Skudlarick, Cho Sung-min, Christina Failma Bachynsky, Candace Chang, Nicole Forrester, Kerry Gershaneck, Justin Goldman, John Hemmings, Hong Seong-ho, Kisuh Jung, Sam Kim, Ariana Lania, Jenny Lin, Matsubara Mihoko, Brooke Mizuno, Liz Morquecho, Cristin Orr Schiffer, Chris Ota, Park Dong-joon, Crystal D. Pryor, David Santoro, Shimizu Aiko, Ariel Stenek, Shenelle Van, Yang Hong Xu, Ting Xu, Adrian Yi, Yang Yi, and Yomon Chisato who I closely worked with during my time in Honolulu as well as the other interns, fellows, staff, and also members of the Young Leader's program. I also thank the Atlantic Council, Center for Strategic and International Studies, Dong-A University, FM Bird, German Marshall Fund of the United States, International Crisis Group, Japan Ground Self-Defense Force, Japan Maritime Self-Defense Force Command and Staff College, Pusan National University, Research Center for Advanced Science and Technology at the University of Tokyo, S. Rajaratnam School of International Studies at Nanyang Technological University, Sasakawa Peace Foundation, University of Malaya, University of Muhammadiyah Malang, US Naval War College, and Waseda University who I have worked for, or collaborated with throughout the years I worked on this project. I also deeply thank the faculty and staff at the Faculty of Asian Studies and the Strategic and Defence

Studies Centre at the Australian National University where I pursued my studies on security affairs and Asian studies, and also the University of New South Wales Canberra School of Humanities and Social Sciences where I wrote my doctoral thesis under the supervision of James Cotton and co-supervisor Aurelia George Mulgan that led to this book. I am also thankful to the Korea Foundation and the Korean Studies Association of Australasia for the scholarship during my studies. The views expressed in this book are mine and do not necessarily reflect the policies or positions of any institutions or persons.

I am also obliged to Akaha Tsuneo, Amako Satoshi, Helmia Asyathri, Patrick Bratton, Jon D. Caverly, Jane Chan Git Yin, Seukhoon Paul Choi, Chun Hong-chan, Zack Cooper, David F. Day, Peter J. Dombrowski, Tonny Dian Effendi, Jay Fidell, Richard Javad Heydarian, Alan Hinge, Ron Huisken, Ikeuchi Satoshi, Ahmad Javid, Harry J. Kazianis, Kim Tae-wan, Kim Hyuk, Koga Kei, Dan Kliman, Koizumi Yu, Chris Lamont, Lee Jin-woo, Lim Suk-jun, Md. Nasrudin Md. Akhir, Mimaki Seiko, Murano Masashi, Murata Aya, Nagakura-Stapf Makiko, Nobe Risa, Tate Nurkin, Oh Mi-yeon, Theoben Jerdan Catindig Orosa, Barry Pavel, Park Hong-suk, Ruli Inayah Ramadhoan, Michael Raska, Jeffrey Robertson, Terence Roehrig, Vina Salviana Darvina Soedarwo, Zoe Stanley-Lockman, Clementine G. Starling, Suh Chung-sok, Takeuchi Toshitaka, Tan See Seng, Chris M. Thomas, Watahiki Yasunori, John Watts, Kelvin Wong, and many other colleagues, including numerous unnamed former and incumbent officials from Australia, China, Japan, Republic of Korea, Singapore, Taiwan, the United Kingdom, and the United States. I also thank the chairs, discussants, panelists, and participants at the various conferences where I presented parts of this study. Big thanks are also due to my great friends (you know who you are) for their friendship, fun, and inspiration throughout the time I worked on this project.

I am obliged to the following libraries for allowing me to access many of the materials used in this study: the Australian National University Menzies Library, Kansai-kan of the National Diet Library, Kyungnam University Institute for Far Eastern Studies Library, National Library of Australia, National Library of Singapore, Pusan National University Library, University of Malaya Library, and the United Nations University Library. I am also indebted to my research assistants, Hong Ji-won, Kim Tae-il, and Lee Kang-hyeok who helped me collect and organize the resources used in the study.

I am honored to have this work published by Routledge, and I am thankful to Simon Bates, Jacy Hui, and ShengBin Tan at the Singapore Office, as well as Sathish Mohan at Apex CoVantage and the copy editors, indexers, and typesetters for their help in editing and producing this book.

I owe my deepest gratitude to my family for all their support throughout the years. Words cannot express how thankful I am for all the help, love, and support I received starting from my life in Saku and Tokyo, to a life enriched with many experiences including my studies, work, and travels in various places around the globe. Finally, I am most thankful to my wife and our two children for being who they are and for all their *aloha* spirit, support, patience, and love that have made my life complete. This work is dedicated to them.

Abbreviations

AEW	Airborne Early Warning
ASCM	Anti-Ship Cruise Missile
ASW	Anti-Submarine Warfare
AWACS	Airborne Warning and Control System
C4ISTAR	Command, Control, Communications, Computers, Intelligence, Surveillance, Target Acquisition, and Reconnaissance
CBRNE	Chemical, Biological, Radiological, Nuclear, and Explosives
CCWPK	Central Committee of the Workers' Party of Korea
CIA	Central Intelligence Agency
CIWS	Close-in Weapon System
CMCWPK	Central Military Commission of the Workers' Party of Korea
CPC	Central People's Committee
DDoS	Distributed Denial of Service
DMZ	Demilitarized Zone
DoS	Denial of Service
DPRK	Democratic People's Republic of Korea
ECM	Electronic Countermeasure
EIW	Electronic Intelligence Warfare
EMP	Electromagnetic Pulse
GPB	General Political Bureau
GSD	General Staff Department
HARTS	Hardened Artillery Sites
HUMINT	Human Intelligence
IAEA	International Atomic Energy Agency
ICBM	Intercontinental Ballistic Missile
ICT	Information and Communication Technology
IMINT	Imagery Intelligence
IRBM	Intermediate-Range Ballistic Missile
ISTAR	Intelligence, Surveillance, Target Acquisition, and Reconnaissance
KPA	Korean People's Army
KPAAF	Korean People's Army Air and Anti-Air Force
KPAN	Korean People's Army Navy
KPASOF	Korean People's Army Special Operation Force

KPASRF	Korean People's Army Strategic Rocket Force
KPISF	Korean People's Internal Security Forces
KPRA	Korean People's Revolutionary Army
KPW	DPRK Won
MAD	Military Affairs Department of the CCWPK
MaRV	Maneuverable Reentry Vehicle
MASINT	Measurement and Signatures Intelligence
MCCCWPK	Military Committee of the CCWPK
MDL	Military Demarcation Line
MEL	Mobile-Erector-Launchers
MIA	Ministry of Internal Affairs
MID	Munitions Industry Department of the CCWPK
MIRV	Multiple Independently Targetable Reentry Vehicle
MND	Ministry of National Defence
MNS	Ministry of National Security
MPAF	Ministry of People's Armed Forces
MPS	Ministry of People's Security
MRBM	Medium-Range Ballistic Missile
MLRS	Multiple Launch Rocket System
MSC	Military Security Command
MSoS	Ministry of Social Security
MSS	Ministry of State Security
NDC	National Defence Commission
NLL	Northern Limit Line
O&M	Operations and Maintenance
OGD	Organization and Guidance Department
ORBAT	Order of Battle
OSINT	Open Source Intelligence
R&D	Research and Development
RGB	Reconnaissance General Bureau
RMTU	Reserve Military Training Unit
ROK	Republic of Korea
RYG	Red Youth Guard
SAC	State Affairs Commission
SCAF	Supreme Commander of the Armed Forces
SEC	Second Economic Committee
SES	Surface Effect Ship
SIGINT	Signals Intelligence
SLBM	Submarine-Launched Ballistic Missile
SLOC	Sea Lines of Communication
SPA	Supreme People's Assembly
SRBM	Short-Range Ballistic Missile
SSB	Ballistic Missile Submarine (Non-nuclear)
SSBN	Ballistic Missile Submarine (Nuclear)
TE	Transporter-Erectors

TEL	Transporter-Erector-Launchers
UAS	Unmanned Aerial Systems
UN	United Nations
US	United States of America
USD	US Dollars
USSR	Union of Soviet Socialist Republics
USV	Unmanned Surface Vehicle
UUV	Unmanned Underwater Vehicle
VSV	Very Slender Vessel
WMD	Weapons of Mass Destruction
WPK	Workers' Party of Korea
WPNK	Workers' Party of North Korea
WPRG	Worker-Peasant Red Guard
WPSK	Workers' Party of South Korea

Note on transliteration

Providing accurate transliteration for Korean names and words is both challenging and confusing due to the various styles that exist. In general, this study transliterates names, policies, publications, terms, and so forth according to the former Republic of Korea (ROK) Ministry of Culture and Tourism's New Romanization System with the exception where there are official or more commonly used forms available in both the Democratic People's Republic of Korea (DPRK) and ROK. In addition, the DPRK employs the United Kingdom style of spelling, thus all official terms and quotes are spelled accordingly where appropriate. For consistency, names of Chinese, Japanese, and Korean persons are written in the traditional order of surname and then the given name. All Korean given names have also been hyphenated to avoid confusion (e.g., Kim Jong-un).

1 Introduction

At the celebration of the 75th anniversary of the Workers' Party of Korea (WPK) on 10 October 2020 at Kim Il Sung Square, Kim Jong-un, the third-generation leader of the Democratic People's Republic of Korea (DPRK or North Korea) gave an emotional speech on the challenges the country faced while vowing to further advance the revolution "toward a fresh victory." Kim Jong-un's speech was then followed by one of the largest displays in the Stalinist state's history, showcasing the various assets of the Korean People's Army (KPA) inventory including artillery, multiple launch rocket systems (MLRS), anti-air missiles, drones, ballistic and cruise missiles, small-arms, and a fly-pass by MiG-29 fighter jets decorated with fancy neon lights. While many observers were intrigued by the display of various new weaponry, the developments are consistent with the defense planning doctrine that has been inherited from the country's founding leader Kim Il-sung.

From the outset of the regime, Kim Il-sung focused on building the armed forces that can autonomously defend the state and help achieve the unification of the Korean peninsula. Yet for the first decade or so, the KPA was only relatively well-armed as a military of a young state. More significant developments came in the 1960s when Kim Il-sung issued the new defense planning doctrine known as the Line of Self-Reliant Defence, calling for the: establishment of a cadre army; modernization of the entire army; arming of the populace; and fortification of the whole country.[1] The Line of Self-Reliant Defence was about autonomously advancing the DPRK's ability not only to defend itself but also to bolster the leadership's absolute command and control over the armed forces. From this time forth, the KPA would undergo significant transformative developments.

Fast-forwarding to 2021, the KPA has grown into a massive force, including the various corps, divisions, and commands of the ground forces, KPA Navy (KPAN), KPA Air and Anti-Air Force (KPAAF), KPA Special Operation Force (KPASOF), and the KPA Strategic Rocket Force (KPASRF). Regarding personnel, the KPA has approximately 1.28 million active personnel, with the addition of 600,000 in the Reserve Military Training Unit (RMTU), 5.7 million in the Worker-Peasant Red Guard (WPRG), 1 million in the Red Youth Guard (RYG), and approximately 189,000 in the Korean People's Internal Security Forces (KPISF).[2] In inventory, the KPA is armed with weapons of mass destruction

(WMD) and now has a collection of various ballistic and cruise missiles including developments in short-range ballistic missiles (SRBM), intermediate-range ballistic missiles (IRBM), medium-range ballistic missiles (MRBM), intercontinental ballistic missiles (ICBM), and submarine-launched ballistic missiles (SLBM). As for conventional capabilities, the KPA has a quantitative assortment of conventional capabilities including artillery and MLRS, armored vehicles and tanks, surface combatants and submarines, tactical aircraft, and a massive special operations force with much of them forward-deployed near the Military Demarcation Line (MDL).

In governance, the armed forces work under a hyper-centralized and politicized multi-dimensional command and control system that has been sculpted by the three generations of the leadership since the state's founding. In defense planning, the Central Military Commission of the WPK (CMCWPK) and the State Affairs Commission (SAC) (formerly the National Defence Commission (NDC)) are in charge of making and implementing decisions while the Supreme Commander of the Armed Forces (SCAF) takes operational command and control – all of which is headed by the incumbent leader Kim Jong-un. Even within the military, the KPA Party Committee system functions as not only a harmonizer but also a check-and-balance mechanism between the field commanders and the political commissars to ensure that the KPA is well under the command and control of the leadership.

The KPA has also become a mysterious monstrosity, based on the regime's policies of concealment, deception, and denial. According to a North Korean encyclopedia, "military secrets" such as activities of units, armaments, capabilities, state of readiness, state of military infrastructures, locations, technologies, information about commanders, and daily lives of personnel are considered to be the "lifelines" of the military's combat capabilities and ability to attain victory.[3] While authoritarian states are notorious for their opaqueness, the DPRK is arguably the most extreme example, where it does not publish defense white papers, detailed budgets, hardware procurements, military strategies, and so forth. Even for internal documents, detailed reports are exclusive to the most senior cadres of the military or the party. The Joseon Inmingun – the internal newspaper of the KPA – makes minimal mentions of technical aspects but is filled with ideological articles on the leadership's efforts to defend the state, the KPA's vital roles in the "Korean revolution," songs and poems, and stories of past and present personnel depicted as examples for their dedication, discipline, and loyalty to the WPK leadership. Even for military parades and publicized drills, much of them are orchestrated theatrics to boost bravado and rally domestic support as well as attracting attention to the capabilities Pyongyang wants the world to see while hiding the complex and difficult realities.

The size of the armed forces combined with the totalitarian, secretive characteristics of the regime, and its combative behavior have led to many misconceptions, misinterpretations, and misunderstandings that have obscured proper understanding of the DPRK and the KPA.

First, it is incorrect to think that the DPRK intended on building a massive military for the sake of it or to take on the world. Rather, the KPA is built as *the* tool for the ultimate aim of unifying the Korean peninsula under Pyongyang's terms as stated in both the constitution and the WPK charter, and the absolute minimal goal of ensuring the Kim dynasty's survival. Thus, there is a great deal of consistency and clarity in the DPRK's defense planning, focusing on strategies and capabilities most essential to deterring and neutralizing the threats against Pyongyang's strategic interests on the Korean peninsula, targeting the United States (US), Republic of Korea (ROK or South Korea), and Japan.

Second, assessments on the KPA's capabilities often end with simplistic conclusions based on comparisons to technologically superior powers such as the US, China, and Russia. While the KPA is certainly armed to the teeth, it is also a bloated force largely stocked with various hardware mostly from the Union of Soviet Socialist Republics (USSR or Soviet) and China during the Cold War, or domestic variants based on those imports. More importantly, the KPA continues to suffer from severe logistical problems that constrain them from effectively mobilizing and operating under war conditions. The causes are clear – the state's limited capacity. Even with the heavy devotion of resources to the military sector and establishment of the monstrous military industry, the DPRK never attained the full autonomous ability to fully modernize or sustain the KPA. That said, it is wrong to conclude that the KPA is completely and harmless. Despite the weaknesses in readiness, the KPA has modernized in its own style that adheres to the DPRK's concept of hybrid warfare and military strategy focusing on: "surprise attacks," "quick and decisive wars," and "mixed tactics."[4] In essence, much comes down to Pyongyang's acknowledgment that it is constrained from full-fledged modernization, making it more sensible to pursue alternative, asymmetric means to deal with technologically superior opponents. Based on the earlier, the DPRK has focused on capabilities most vital to meet its strategic and tactical demands, while also utilizing the marginal assets for asymmetric effect.

Third, while the DPRK's military actions have often been abrupt, it is another to say that they are irrational, reckless, trigger-happy, and unpredictable. Certainly, the DPRK is one of the evilest regimes in history, being an encyclopedia of human rights abuses, master of extortion and manipulation in international affairs, and an actor that relies on violence to achieve its aims. Yet it is important to distinguish morality and rationality. The truth that the DPRK is not irrational – as they would have disappeared long ago if they were. Rather, the DPRK understands that every strategic decision and action would make or break the survival of the regime. Such characteristics explain why the DPRK has defied the odds, preventing regime collapse and undertaking two power successions despite the economic failures, as well as surviving the brinkmanship and confrontations with much more powerful states.

Fourth, relating to the third, it is a mistake to assume that the DPRK has not changed. Indeed, there are many aspects such as the grand strategic goals, prioritization of the military, and unconditional priority to preserve the Kim dynasty that remain unmolested. Yet key changes and developments are evident. In the leadership, while adhering to the same policies and visions, the three generations of the Kim dynasty have practiced different styles of leadership, employed different political tactics, and even reconfigured the state governance institutions to ensure the regime's survival. As for the society, due to the continued widespread poverty (particularly in the countryside), the old socialist state-controlled economic practices have withered, with market activities becoming the new norm. Then for the military, although the KPA still maintains the massed forces with strategies and doctrines inherited from Kim Il-sung, they have nonetheless significantly modernized since the 1960s with the acquisition of strategic weapons and some upgrades to the conventional inventory. The bottom line is that the DPRK has modernized and transformed without reform. Although the outcomes have certainly not been pleasing from the socio-economic, human rights, and international security viewpoint, the DPRK has continuously adapted to the changing circumstances and has been innovative in its measures to survive and achieve their strategic goals.

The misconceptions, misinterpretations, and misunderstandings often come from the simplistic characterization of the DPRK's strategies, the KPA's capabilities, as well as the Stalinist state's behavior. Yet the "take it as it is" approach without proper context will have major consequences. Regardless of the state of the KPA, the DPRK continues to pose a major threat to international security. In the most extreme case, the DPRK could conduct a full-scale attack against the ROK in an attempt either to unify the Korean peninsula or to neutralize the threat they face from the ROK and US. Such campaigns would also involve strikes against not only Japan but also the US assets in the Pacific and even the mainland. On the other end of the scale, the DPRK would continue to use the threat of force, as well as conducting a variety of provocations and small-scale attacks that would embolden the "gray-zone" conflict and destabilize the Indo–Pacific region (aptly termed by Cha as "hit and run" tactics).[5] In any case, there is a great deal of uncertainty should conflict break out. A war of any scale with the DPRK will have significant ramifications on the region, with immediate loss of lives on both sides of the MDL, potential volley of WMD, and most likely intervention by China and potentially Russia. In addition, there are also questions over the regime's future, where there is the risk of internal instability and even a potential civil war within North Korea that will open a new range of contingencies. Given the risks and threats faced, it is imperative to properly understand how the structures, processes, and circumstances have shaped the DPRK's military readiness. Doing so not only will help make sense of the developments but also in formulating effective strategies to adequately deal with the threat and avert crises.

Understanding defense planning and military readiness

Strategy has been broadly defined as setting the means to achieve the state's ends.[6] The critical part in setting and pursuing the strategy is defense planning to generate the optimal military readiness to achieve the state's strategic aims and objectives. Gray terms defense planning as the "preparations for the defense of a polity in the future."[7] Layne provides a sound description of defense planning, calling it as a three-step process that involves "determining a state's vital security interests; identifying the threats to those interests; and deciding how best to employ the state's political, military, and economic resources to protect those interests."[8] Defense planning is heavily connected to systems analysis, defined by Quade as "[placing] each element in its proper context so that in the end the system as a whole may achieve its aims with a minimal expenditure of resources."[9] Scholars such as Liotta and Lloyd have also tried to map out the strategic policy-making process, creating a flowchart of questions decision-makers face in formulating and planning their strategies.[10]

However, although we can logically understand what defense planning is for, and sketch out the framework of the process, much focus is needed on military readiness that renders the state's ability to effectively and efficiently utilize the capabilities to achieve the state's strategic aims and objectives. The best discussions on readiness are offered by Betts who talked about the critical importance of balancing "structural" and "operational" readiness, as well as three key questions in attaining the optimal balance, being readiness: "for what," "of what," and "for when."[11] The problem is how states generate and manage their military readiness. While states may have a blueprint, the defense planning process is inevitably affected by a range of political, economic, social, and environmental factors that often act as constraints. Moreover, the state of readiness would be determined by a variety of factors other than weapons systems and infrastructures, including administrative and bureaucratic competence, education and training, experience, logistics, quality of personnel, supplies, and so forth.[12] Even in operations, context such as politics, military, economy, society, information, infrastructure, physical environment, and time (PMESII-PT) is vital.[13] Thus, the focus on readiness also reminds us that defense planning would inevitably involve and is affected by a significant variety of factors, requiring a more holistic approach.

Much context is needed regarding the endogenous and exogenous factors. Despite some scholarly differences, there is a general consensus that external factors give hints to *what* a state will react to, whereas internal factors shed light on *how* they will react and build their capabilities.[14] Exogenous factors, therefore, relate much more with the input and output factors, shaping perceptions and justifying the government's strategy while also revealing the actual effects of the strategies and actions they employ. Yet the actual processing phase in defense planning is shaped by and works in accordance with the state's particular set of structures, processes, and circumstances. The influencing factors are not only political, economic, and military but also include culture, environment, geography, and society.[15]

Capacity is arguably the biggest constraint, as it is one thing for a state to identify what it *wants* to do; but whether it *can* do so is another. The biggest capacity factors are economic and industrial, determining not only the production and acquisition of systems but also the logistical items that are vital to operating the assets. Knorr correctly argues that "Whatever the structure of the economy, every society faces the same basic questions in regard to production, distribution, budgeting and management of technology."[16] While certainly true, Knorr's argument reminds us about how we need to closely examine the specific characteristics of states to investigate how those questions are addressed.

Obviously, one vital area is military expenditures that indicate the amount of resources to strengthen and sustain military readiness. Huisken broadly describes the contents of military expenditures, including: pay, allowances, and pension for serving and retired personnel; operations and maintenance; procurement; research and development (R&D); construction; military aid; civil defense; paramilitary forces; and military aspects of joint civilian–military projects (e.g., space or atomic energy).[17] However, states have their own ways of defining military expenditures, and there are also differences in the sourcing of the funds, accounting systems, costs, purchasing power, and also currency values. More importantly, the amount of resources devoted to the military sector does not necessarily render the full state of readiness. Even if a state devotes a large share of its resources to the military sector, whether that is sufficient or not much depends on how the resources are spent.[18] Much comes down to the balanced distribution of resources to two categories: one-off investments for R&D and procurements; and recurring operational costs for operations and maintenance (O&M).[19]

Industrial capacity is also important, particularly for states who seek to autonomously enhance and sustain their military readiness. While there are states such as the US, USSR/Russia, and China that constructed mega military–industrial complexes, states are seldom fully self-reliant or sufficient. Having a productive and sustainable military–industrial complex much depends on high levels of sufficiency in resources, R&D, human resources, and a sound economic-industrial ecosystem.

The various stimuli cause dilemmas that acutely affect the defense planning process and consequently the military's readiness. Regardless of regime types, the fundamental management dilemmas are shared, including: agenda-setting, "guns and butter," resource allocation in the military, cost-effectiveness, balance of force structural and operational readiness, political debates and implications, resource management, the security dilemma, and others.[20] Yet the problem is that rational decisions to deal with the dilemmas are regularly undermined by politics.[21] Murray and Grimsley identify politics, ideology, and geography as the core components of a state's strategic culture, which "may make it difficult for a state to evolve sensible and realistic approaches to the strategic problems that confront it."[22] Snyder also argues that "'decision-making' will be a process of rationalization rather than that of rationality," where decisions are made, or at least influenced by factors such as incentives, biases, and doctrinal oversimplifications.[23] Richardson also argues that states are often compelled to strengthen their militaries as "they follow their traditions which are fixtures and their instincts which are mechanical."[24] Gray also offers an important point, arguing that even though defense planning is about preparing for future uncertainties, "in practice it cannot help but be dominated by people whose whole knowledge and experience is only of the past."[25] Another essential aspect is the styles of governance and bureaucratic processes.[26] Biddle criticizes material-centric analysis that excludes non-material causatives such as administrative competence and skill that impact military effectiveness.[27] The level of administrative competence could be adversely affected by extreme levels of centralization and politicization, leading to decisions and processes based on regime security as opposed to best practice. Despite the varying approaches, the arguments point to the general direction where decisions and processes are often politicized or affected by mindsets that undermine rational decision-making.

The significance of politics in defense planning raises important questions over whether particular types of political systems lead to particular decisions. Scholars such as Bathurst have extensively explained how a nation's military strategies and strategic perceptions do reflect its cultural, political, and social characteristics.[28] While convincing, there are caveats, as both democratic and authoritarian states use security for political purposes and use fear to justify defense planning decisions.[29] Rather, more attention is needed on the process. In authoritarian states where there is a centralized hierarchy that places the leader or the ruling party above the state and its laws, decisions are made and processed with greater cohesion and speed.[30] At the same time, centralized systems lead to self-imposed path-dependence where leaders scrutinize and eradicate ideas that may in fact be rational from the military readiness viewpoint but unacceptable for their potential to destabilize or weaken the ruler's authority.

Clarity is also needed on the instinctive factors that influence defense planning, where there are often confusions between ideologies and strategic culture. Much care is needed when talking about ideologies, as they are politically constructed quasi-religious ethics that are used to legitimize the regime's visions, decisions, and actions. Rather, the more compelling factors are derived from the state's political cultures that lead to distinct perceptions and decisions.[31] Johnston distinguishes strategic culture from ideology, arguing that the former is a set of instinctive factors that are "rooted in the early or formative experiences of the state, and influenced to some degree by the philosophical, political, cultural, and cognitive characteristics of the state and its elites."[32] Thus to understand the psyche of states, one must not merely look at political systems and their ideologies but also historical experiences, national identity, and geographical factors that construct perceptions and behavior.

Even when states are confident in their decisions, politicized decisions lead to plans-reality mismatches that ironically undermine the state's ability to achieve its strategic aims and objectives.[33] On the one hand, under-readiness would make the military unable to achieve the state's aims and objectives, while, on the other hand, over-readiness would up the security dilemma and could also trigger fatigues.[34] Indeed, measuring and striking the optimal level of readiness is no easy task, particularly as the actual effectiveness of military readiness pivots on the readiness of the opponent and the nature of the conflict. While it is easy for us to know (largely in hindsight) when states are not ready, it is hard to know when they are. In cases where military readiness is insufficient, states will need to either go back to the drawing board or devise ways to make the most of the assets they have, such as innovation in operational and tactical concepts and doctrines as well as strategic maneuvers.

While broad, the discussions thus far point to the importance of the country-specifics. How the governments manage their military's readiness and deal with the decision-making dilemmas are largely determined by interests and capacities that are shaped by their specific structures, processes, and circumstances. Still, balance is critical. As Russett warns, one must be sure to neither ignore the country-specifics nor go to other extremes of assuming that the uniqueness of states exempts them from general patterns.[35] Thus, regardless of the regime's nature, they would still face fundamental questions like other states, while the uniqueness of states would render the particular dilemmas and questions they face and how they address them, consequently leading to their state of military readiness.

The conceptual debates in defense planning provide key questions in better analyzing the DPRK's management of the KPA's readiness. Broadly, the major works on the DPRK's military strategies and the armed forces to date fall into six categories. The first category includes historical analysis of the KPA.[36] The second category consists of studies on the governance and political aspects, such as the command and control hierarchy, the decision-making processes, institutional structures, and party-military relations.[37] The third category comprises of economic analyses, including the military budget, military industry, economic policies that prioritize the military than state development, and so forth.[38] The fourth category consists of analyses of the technical aspects such as the KPA's assets and order of battle (ORBAT).[39] The fifth category, which sometimes overlaps with the fourth, includes analyses on the DPRK's military strategies and tactics and outlines the likely courses of action.[40] The sixth category consists of empirical analysis on the DPRK's patterns of strategic behavior and interactions.[41] The aforementioned analyses are informative and certainly have offered much to the debates on the North Korean armed forces and their strategies, and do provide valuable discussions that are relevant to understand Pyongyang's defense planning – particularly, despite the shortage of credible data. At the same time, there is a major gap that must be filled to understand the DPRK's defense planning process itself, and the readiness of the KPA. Indeed, there are some comprehensive studies on the KPA that do come close to sketching out the DPRK's defense planning and military readiness.[42] Yet still, there remains to be a niche to conceptually and contextually connect the dots to understand how not only the DPRK's strategies but also the political, economic, social, and environmental structures, processes, and circumstances have shaped its military readiness, as well as their implications.

Assessing the DPRK's military readiness

The aim of this study is to analyze the DPRK's defense planning and its impact on the KPA's readiness. This study does not look deep into the DPRK's strategic behavior, nor the action–reaction interactions with other states as these topics have already been well covered by various studies. Rather, this study focuses on how the DPRK developed and managed its armed forces to achieve the state's strategic aims and objectives. The study is based on five key points. The first illustrates how the combination of circumstantial developments and sense of national and political identity have molded the DPRK's pursuit of its strategic goals, with particular focus on self-reliance and offensive strategies. The second explains how the political and economic structures, processes, and circumstances have created a set of defense planning dilemmas that compel the DPRK to make trade-offs not only between interests and capacity but also regime security and military readiness. The third illustrates how the DPRK's defense planning has significantly impacted the military readiness of the KPA, creating issues both in force structural readiness and operational readiness. The fourth explains how despite the problems in readiness, the DPRK's warfighting concepts and doctrines as well as the patterns in modernization still present acute threats. The fifth explains how the centralized and politicized command and control of the armed forces would show its reverse effects should any significant instability unfold in the regime, potentially leading to the military's fragmentation and worse, a multilateral civil war within North Korea.

The following six chapters are structured to provide comprehensive and contextualized understanding of the DPRK's defense planning and its impact on the KPA's readiness. Chapter 2 will overview the circumstantial developments and the nature of the regime that shaped the DPRK's strategic mindset and threat perceptions. Chapter 3 examines the DPRK's command and control of defense planning and the armed forces, revealing the centralized and politicized mechanism designed to seamlessly execute the leadership's orders while keeping the KPA loyal. Chapter 4 looks at the DPRK's economic and industrial capacity, highlighting the key constraints despite Pyongyang's quest to attain autonomous capacity to strengthen the KPA. Chapter 5 examines the defense planning framework, evidencing how the strategic, political, and economic factors have created major dilemmas for Pyongyang in pursuing the Line of Self-Reliant Defence to strengthen its military readiness. Chapter 6 will then assess the readiness of the KPA, looking at both the force structural and operational readiness aspects. Finally, Chapter 7 will outline the implications of the study, looking at the military threat, proliferation concerns, and also the uncertain fate of the regime, and will end with a discussion on the problems in the strategies vis-à-vis the DPRK.

Notes

1 The DPRK often uses the terms Jawijeok Gunsa Roseon (Line of Self-Reliant Defence) and the Dangeui Gunsa Roseon (Military Lines of the Party). However, foreign media and writings often refer to it as the Sadae Gunsa Roseon (Four Grand Military Lines or Four-Point Military Guidelines).

2 International Institute for Strategic Studies, *The Military Balance 2020* (London, UK: International Institute for Strategic Studies, 2020), 284–86; ROK Ministry of National Defense, *Defense White Paper 2018* (Seoul, ROK: ROK Ministry of National Defense, 2018), 35.

3 Baekgwasajeon Chulpansa, *joseon daebaekgwasajeon [Korea Encyclopedia]*, vol. 3 (Pyongyang, DPRK: Baekgwasajeon Chulpansa, 1995), 272.

4 US Department of the Army, *ATP 7-100.2: North Korean Tactics* (Washington, DC: US Department of the Army, July 2020), 1.13. The US Department of the Army defines hybrid warfare as "the diverse and dynamic combination of regular forces, irregular forces, terrorist forces, or criminal elements unified to achieve mutually benefitting threat effects."; US Department of the Army, *ADRP 3–0: Operations* (Washington, DC: US Department of the Army, October 2017), 1–5.
5 Victor D. Cha, *Testimony Before the U.S. House of Representatives, Committee on Foreign Affairs* (Washington, DC: Center for Strategic and International Studies, 10 March 2011), 5.
6 B. H. Liddell Hart, *Strategy*, 2nd ed. (New York: Praeger, 1967), 335; Lawrence Freedman, "Strategic Coercion," in *Strategic Coercion: Concepts and Cases*, ed. Lawrence Freedman (Oxford, UK: Oxford University Press, 1998), 15.
7 Colin S. Gray, *Strategy and Defence Planning: Meeting the Challenge of Uncertainty* (New York, NY: Oxford University Press, 2014), 4.
8 Christopher Layne, "From Preponderance to Offshore Balancing: America's Future Grand Strategy," *International Security* 22, no. 1 (1997), 88.
9 E. S. Quade, "Introduction," in *Analysis for Military Decisions*, ed. E. S. Quade (Chicago, IL: Rand McNally, 1964), 4.
10 P. H. Liotta and Richmond M. Lloyd, "From Here to There: The Strategy and Force Planning Framework," *Naval War College Review* 58, no. 2 (2005).
11 Richard K. Betts, *Military Readiness: Concepts, Choices, Consequences* (Washington, DC: Brookings Institution Press, 1995), 33, 40–43.
12 For example, the US Department of Defense stresses the importance of doctrine, organization, training, materiel, leadership and education, personnel, and facilities (DOTMLPF).
13 See: Ryan Burke, "Operational Design," in *Military Strategy, Joint Operations, and Airpower: An Introduction*, ed. Ryan Burke, Michael W. Fowler, and Kevin McCaskey (Washington, DC: Georgetown University Press, 2018), 34–37; Jeffrey M. Reilly, *Operational Design: Distilling Clarity from Complexity for Decisive Action* (Maxwell Air Force Base, AL: Air University Press, 2012), 5; US Joint Chiefs of Staff, *Joint Publication 3–0: Joint Operations* (Washington, DC: US Joint Chiefs of Staff, 17 January 2017), IV.3.
14 For example, clearly defined exogenous factors smooth the military organizational process. See: Risa A. Brooks, "Conclusion," in *Creating Military Power: The Sources of Military Effectiveness*, ed. Risa A. Brooks and Elizabeth A. Stanley (Stanford, CA: Stanford University Press, 2007), 229.
15 See: Risa A. Brooks, "Introduction: The Impact of Culture, Society, Institutions and International Forces on Military Effectiveness," in *Creating Military Power: The Sources of Military Effectiveness*, ed. Risa A. Brooks and Elizabeth A. Stanley. Stanford, CA: Stanford University Press, 2007), 1-22.
16 Klaus Knorr, *Military Power and Potential* (Lexington, MA: Heath Lexington Books, 1970), 114.
17 Ronald Huisken, "The Meaning and Measurement of Military Expenditure," in *Stockholm International Peace Research Institute Research Report;* no. 10 (Stockholm, Sweden: Stockholm International Peace Research Institute, 1973), 7.
18 Michael E. O'Hanlon, *The Science of War: Defense Budgeting, Military Technology, Logistics, and Combat Outcomes* (Princeton, NJ: Princeton University Press, 2009), 19.
19 See: N. V. Breckner and J. W. Noah, "Costing of Systems," in *Defense Management*, ed. Stephen Enke (Englewood Cliffs, NJ: Prentice-Hall, 1967), 49; Betts, *Military Readiness: Concepts, Choices, Consequences*, 35-62.
20 For discussions on defense planning dilemmas, see: Colin S. Gray, "Coping with Uncertainty: Dilemmas of Defense Planning," *Comparative Strategy* 27, no. 4

(2008); Dennis M. Drew and Donald M. Snow, *Making Twenty-First-Century Strategy: An Introduction to Modern National Security Processes and Problems* (Maxwell Air Force Base, AL: Air University Press, 2006); Alan Hinge, *Australian Defence Preparedness: Principles, Problems and Prospects: Introducing Repertoire of Missions (ROMINS) a Practical Path to Australian Defence Preparedness* (Canberra, Australia: Australian Defence Studies Centre, 2000).

21 These points are thoroughly explained by scholars such as Knorr and Gray. See: Knorr, *Military Power and Potential*; Gray, *Strategy and Defence Planning: Meeting the Challenge of Uncertainty.*

22 Williamson Murray and Mark Grimsley, "Introduction: On Strategy," in *The Making of Strategy: Rulers, States, and War*, ed. Williamson Murray, MacGregor Knox, and Alvin H. Bernstein (Cambridge, UK and New York: Cambridge University Press, 1994), 3.

23 Jack L. Snyder, *The Ideology of the Offensive: Military Decision Making and the Disasters of 1914* (Ithaca, NY: Cornell University Press, 1984), 18–19, 32.

24 Lewis Fry Richardson, *Arms and Insecurity: A Mathematical Study of the Causes and Origins of War*, ed. Nicolas Rashevsky and Ernesto Trucco (Ann Arbor, MI: Boxwood Press, 1978), 12–13.

25 Gray, *Strategy and Defence Planning: Meeting the Challenge of Uncertainty*, 19.

26 For through discussions on this topic, see: Graham T. Allison, *Essence of Decision: Explaining the Cuban Missile Crisis* (Boston, MA: Little, Brown, 1971).

27 For example, Biddle argues, "[T]hreat assessments based on the numbers and types of hostile weapons are likely to overestimate the real capabilities of enemies with modern equipment but limited skills, and underestimate militaries with older equipment but high skills." Stephen Biddle, "Explaining Military Outcomes," in *Creating Military Power: The Sources of Military Effectiveness*, ed. Risa A. Brooks and Elizabeth A. Stanley (Stanford, CA: Stanford University Press, 2007), 218.

28 Robert B. Bathurst, *Intelligence and the Mirror: On Creating an Enemy* (Oslo, Norway and London, UK: International Peace Research Institute; Sage Publications, 1993).

29 See: Gray, *Strategy and Defence Planning: Meeting the Challenge of Uncertainty*, 143, 182.

30 For discussions on cohesion, see: Ashley J. Tellis, *Measuring National Power in the Postindustrial Age* (Santa Monica, CA: RAND/Arroyo Center, 2000), 109–11.

31 For example, Thakur argues, "[States] adopt a military strategy different from choices made by other countries . . . even when confronted with essentially similar facts." Ramesh Thakur, "New Zealand: The Security and Tyranny of Isolation," in *Strategic Cultures in the Asia-Pacific Region*, ed. Ken Booth and Russell B. Trood (New York, NY: St. Martin's Press, 1999), 314.

32 Alastair Iain Johnston, "Thinking about Strategic Culture," *International Security* 19, no. 4 (1995): 34.

33 Gray, *Strategy and Defence Planning: Meeting the Challenge of Uncertainty*, 140–41. Also see: Franklin C. Spinney and James Clay Thompson, *Defense Facts of Life: The Plans/Reality Mismatch* (Boulder, CO: Westview Press, 1984).

34 Betts, *Military Readiness: Concepts, Choices, Consequences*, 28–30.

35 Bruce Russett, "International Interactions and Processes: The Internal vs. External Debate Revisited," in *Political Science: The State of the Discipline*, ed. Ada W. Finifter and American Political Science Association (Washington, DC: American Political Science Association, 1983), 561.

36 Studies include: Kwang-soo Kim, "joseoninmingun changseolgwa baljeon [The Establishment and Development of the Korean People's Army]," in *bukhangunsamunjeeui jaejomyeong [The Military of North Korea: A New Look]* (Paju, ROK: Hanul Academy, 2006); Katsuichi Tsukamoto, *kitachousengunto seiji [The North*

Korean Army and Politics] (Tokyo, Japan: Hara Shobo, 2000); Robert A. Scalapino and Chong-sik Lee, *Communism in Korea* (Berkeley, CA: University of California Press, 1972); Wan-gyu Choi, "joseoninminguneui hyeongseonggwa baljeon [The Formation and Development of the KPA]," in *bukhaneui gunsa [North Korean Military Affairs]*, ed. Bukhan Yeongu Hakhoi (Seoul, ROK: Gyeongin Munhwasa, 2006).

37 Studies include: Chin-moo Kim, "bukhaneui jeongchaek gyeoljeongeseo gunbueui yeonghyang [The Military's Influence in North Korea's Policy Decisions]," ibid.; Yeong-hoon Lee, *bukhaneul umjikineun him: gunbueui paegwon gyeongjaeng [The Forces that Move North Korea: The Competition for Hegemony in the Military]* (Paju, ROK: Salim Books, 2012); Jae-hong Ko, *Kim Jong-il chejeeui bukhangun yeongu [Study of the North Korean Military under the Kim Jong-il Regime]* (Seoul, ROK: Institute for National Security Strategy, 2011); Dae-keun Yi, *bukhanguneun woae kudetareul haji ana [Why Don't the Korean People's Army Make a Coup]* (Paju, ROK: Hanul Academy, 2003).

38 Studies include: Sung-bin Choi, Jae-moon Yoo, and Si-woo Kwak, *bukhan gunsusaneob gaehwang [Current State of the North Korean Military Industry]* (Seoul, ROK: Korea Institute for Defense Analyses, 2005); Taik-young Hamm, *Arming the Two Koreas: State, Capital and Military Power* (London, UK and New York: Routledge, 1999); Kang-taeg Lim, "bukhaneui gunsusaneob jeongchaek [North Korea's Military Industry Policies]," in *bukhaneui gunsa [North Korean Military Affairs]*, ed. Bukhan Yeongu Hakhoi (Seoul, ROK: Gyeongin Munhwasa, 2006); Chae-gi Sung, "bukhan gongpyogunsabi silchee daehan jeongmil jaebunseok [Detailed Re-examination of North Korea's Official Military Expenditures]," in *bukhaneui gunsa [North Korean Military Affairs]*, ed. Bukhan Yeongu Hakhoi (Seoul, ROK: Gyeongin Munhwasa, 2006); Chae-gi Sung, Joo-hyun Park, Jae-ok Park, and O-bong Kwon, *bukhan gyeongjewigi 10nyeongwa gunbijeunggang neungryeok [North Korea's Decade of Economic Crisis and Capacity for Military Buildup]* (Seoul, ROK: Korea Institute for Defense Analyses Press, 2003).

39 Studies include: Joseph S. Bermudez Jr., *The Armed Forces of North Korea* (St. Leonards, Australia: Allen & Unwin, 2001); Anthony H. Cordesman, *The Military Balance in the Koreas and Northeast Asia* (Washington, DC: Center for Strategic and International Studies, 2017); Andrew Scobell and John M. Sanford, *North Korea's Military Threat: Pyongyang's Conventional Forces, Weapons of Mass Destruction, and Ballistic Missiles* (Carlisle, PA: US Army War College Strategic Studies Institute, 2007).

40 Studies include: US Department of the Army, *ATP 7–100.2: North Korean Tactics*; James M. Minnich, *The North Korean People's Army: Origins and Current Tactics* (Annapolis, MD: Naval Institute Press, 2005).

41 Studies include: Narushige Michishita, *North Korea's Military-Diplomatic Campaigns, 1966–2008* (London, UK and New York: Routledge, 2010); Van Jackson, *Rival Reputations: Coercion and Credibility in US-North Korea Relations* (Cambridge, UK: Cambridge University Press, 2016).

42 Studies include: Yang-ju Kwon, *bukhangunsaeui ihae [The Comprehension of North Korean Military]*, Expanded ed. (Seoul, ROK: Korea Institute of Defense Analyses, 2014); Yong-won Yoo, Beom-chul Shin, and Jin-a Kim, eds., *bukhangun sikeurit ripoteu [North Korea Military Secret Report]* (Seoul, ROK: Planet Media, 2013); ROK Military Academy, *bukhanhak [North Korea Studies]* (Seoul, ROK: Hwanggeumal, 2006).

References

Allison, Graham T. *Essence of Decision: Explaining the Cuban Missile Crisis.* Boston, MA: Little, Brown, 1971.

Baekgwasajeon Chulpansa. *joseon daebaekgwasajeon [Korea Encyclopedia].* Vol. 3. Pyongyang, DPRK: Baekgwasajeon Chulpansa, 1995.

Bathurst, Robert B. *Intelligence and the Mirror: On Creating an Enemy.* Oslo, Norway and London, UK: International Peace Research Institute and Sage Publications, 1993.

Bermudez Jr., Joseph S. *The Armed Forces of North Korea.* St. Leonards, Australia: Allen & Unwin, 2001.

Betts, Richard K. *Military Readiness: Concepts, Choices, Consequences.* Washington, DC: Brookings Institution Press, 1995.

Biddle, Stephen. "Explaining Military Outcomes." In *Creating Military Power: The Sources of Military Effectiveness*, edited by Risa A. Brooks and Elizabeth A. Stanley. Stanford, CA: Stanford University Press, 2007.

Breckner, N. V., and J. W. Noah. "Costing of Systems." In *Defense Management*, edited by Stephen Enke. Englewood Cliffs, NJ: Prentice-Hall, 1967.

Brooks, Risa A. "Conclusion." In *Creating Military Power: The Sources of Military Effectiveness*, edited by Risa A. Brooks and Elizabeth A. Stanley, 252. Stanford, CA: Stanford University Press, 2007a.

———. "Introduction: The Impact of Culture, Society, Institutions and International Forces on Military Effectiveness." In *Creating Military Power: The Sources of Military Effectiveness*, edited by Risa A. Brooks and Elizabeth A. Stanley. Stanford, CA: Stanford University Press, 2007b.

Burke, Ryan. "Operational Design." In *Military Strategy, Joint Operations, and Airpower: An Introduction*, edited by Ryan Burke, Michael W. Fowler, and Kevin McCaskey. Washington, DC: Georgetown University Press, 2018.

Cha, Victor D. *Testimony Before the U.S. House of Representatives, Committee on Foreign Affairs.* Washington, DC: Center for Strategic and International Studies, 10 March 2011.

Choi, Sung-bin, Jae-moon Yoo, and Si-woo Kwak. *bukhan gunsusaneob gaehwang [Current State of the North Korean Military Industry].* Seoul, ROK: Korea Institute for Defense Analyses, 2005.

Choi, Wan-gyu. "joseoninminguneui hyeongseonggwa baljeon [The Formation and Development of the KPA]." In *bukhaneui gunsa [North Korean Military Affairs]*, edited by Bukhan Yeongu Hakhoi. Seoul, ROK: Gyeongin Munhwasa, 2006.

Cordesman, Anthony H. *The Military Balance in the Koreas and Northeast Asia.* Washington, DC: Center for Strategic and International Studies, 2017.

Drew, Dennis M., and Donald M. Snow. *Making Twenty-First-Century Strategy: An Introduction to Modern National Security Processes and Problems.* Maxwell Air Force Base, AL: Air University Press, 2006.

Freedman, Lawrence. "Strategic Coercion." In *Strategic Coercion: Concepts and Cases*, edited by Lawrence Freedman. Oxford, UK: Oxford University Press, 1998.

Gray, Colin S. "Coping with Uncertainty: Dilemmas of Defense Planning." *Comparative Strategy* 27, no. 4 (2008).

———. *Strategy and Defence Planning: Meeting the Challenge of Uncertainty.* New York, NY: Oxford University Press, 2014.

Hamm, Taik-young. *Arming the Two Koreas: State, Capital and Military Power.* London, UK and New York, NY: Routledge, 1999.

Hart, B. H. Liddell. *Strategy.* 2nd ed. New York, NY: Praeger, 1967.

Hinge, Alan. *Australian Defence Preparedness: Principles, Problems and Prospects: Introducing Repertoire of Missions (ROMINS) a Practical Path to Australian*

Defence Preparedness. Canberra, Australia: Australian Defence Studies Centre, 2000.

Huisken, Ronald. "The Meaning and Measurement of Military Expenditure." In *Stockholm International Peace Research Institute Research Report.* Stockholm, Sweden: Stockholm International Peace Research Institute, 1973.

International Institute for Strategic Studies. *The Military Balance 2020.* London, UK: International Institute for Strategic Studies, 2020.

Jackson, Van. *Rival Reputations: Coercion and Credibility in US-North Korea Relations.* Cambridge, UK: Cambridge University Press, 2016.

Johnston, Alastair Iain. "Thinking about Strategic Culture." *International Security* 19, no. 4 (1995).

Kim, Chin-moo. "bukhaneui jeongchaek gyeoljeongeseo gunbueui yeonghyang [The Military's Influence in North Korea's Policy Decisions]." In *bukhaneui gunsa [North Korean Military Affairs]*, edited by Bukhan Yeongu Hakhoi. Seoul, ROK: Gyeongin Munhwasa, 2006.

Kim, Kwang-soo. "joseoninmingun changseolgwa baljeon [The Establishment and Development of the Korean People's Army]." In *bukhangunsamunjeeui jaejomyeong [The Military of North Korea: A New Look].* Paju, ROK: Hanul Academy, 2006.

Knorr, Klaus. *Military Power and Potential.* Lexington, MA: Heath Lexington Books, 1970.

Ko, Jae-hong. *Kim Jong-il chejeeui bukhangun yeongu [Study of the North Korean Military under the Kim Jong-il Regime].* Seoul, ROK: Institute for National Security Strategy, 2011.

Kwon, Yang-ju. *bukhangunsaeui ihae [The Comprehension of North Korean Military].* Expanded ed. Seoul, ROK: Korea Institute of Defense Analyses, 2014.

Layne, Christopher. "From Preponderance to Offshore Balancing: America's Future Grand Strategy." *International Security* 22, no. 1 (1997).

Lee, Yeong-hoon. *bukhaneul umjikineun him: gunbueui paegwon gyeongjaeng [The Forces that Move North Korea: The Competition for Hegemony in the Military].* Paju, ROK: Salim Books, 2012.

Lim, Kang-taeg. "bukhaneui gunsusaneob jeongchaek [North Korea's Military Industry Policies]." In *bukhaneui gunsa [North Korean Military Affairs]*, edited by Bukhan Yeongu Hakhoi. Seoul, ROK: Gyeongin Munhwasa, 2006.

Liotta, P. H., and Richmond M. Lloyd. "From Here to There: The Strategy and Force Planning Framework." *Naval War College Review* 58, no. 2 (2005).

Michishita, Narushige. *North Korea's Military-Diplomatic Campaigns, 1966–2008.* London, UK and New York, NY: Routledge, 2010.

Minnich, James M. *The North Korean People's Army: Origins and Current Tactics.* Annapolis, MD: Naval Institute Press, 2005.

Murray, Williamson, and Mark Grimsley. "Introduction: On Strategy." In *The Making of Strategy: Rulers, States, and War*, edited by Williamson Murray, MacGregor Knox, and Alvin H. Bernstein. Cambridge, UK and New York, NY: Cambridge University Press, 1994.

O'Hanlon, Michael E. *The Science of War: Defense Budgeting, Military Technology, Logistics, and Combat Outcomes.* Princeton, NJ: Princeton University Press, 2009.

Quade, E. S. "Introduction." In *Analysis for Military Decisions*, edited by E. S. Quade. Chicago, IL: Rand McNally, 1964.

Reilly, Jeffrey M. *Operational Design: Distilling Clarity from Complexity for Decisive Action*. Maxwell Air Force Base, AL: Air University Press, 2012.

Richardson, Lewis Fry. *Arms and Insecurity: A Mathematical Study of the Causes and Origins of War*. Edited by Nicolas Rashevsky and Ernesto Trucco. Ann Arbor, MI: Boxwood Press, 1978.

ROK Military Academy. *bukhanhak [North Korea Studies]*. Seoul, ROK: Hwanggeumal, 2006.

ROK Ministry of National Defense. *Defense White Paper 2018*. Seoul, ROK: ROK Ministry of National Defense, 2018.

Russett, Bruce. "International Interactions and Processes: The Internal vs. External Debate Revisited." In *Political Science: The State of the Discipline*, edited by Ada W. Finifter and American Political Science Association. Washington, DC: American Political Science Association, 1983.

Scalapino, Robert A., and Chong-sik Lee. *Communism in Korea*. Berkeley, CA: University of California Press, 1972.

Scobell, Andrew, and John M. Sanford. *North Korea's Military Threat: Pyongyang's Conventional Forces, Weapons of Mass Destruction, and Ballistic Missiles*. Carlisle, PA: US Army War College Strategic Studies Institute, 2007.

Snyder, Jack L. *The Ideology of the Offensive: Military Decision Making and the Disasters of 1914*. Ithaca, NY: Cornell University Press, 1984.

Spinney, Franklin C., and James Clay Thompson. *Defense Facts of Life: The Plans/Reality Mismatch*. Boulder, CO: Westview Press, 1984.

Sung, Chae-gi. "bukhan gongpyogunsabi silchee daehan jeongmil jaebunseok [Detailed Re-examination of North Korea's Official Military Expenditures]." In *bukhaneui gunsa [North Korean Military Affairs]*, edited by Bukhan Yeongu Hakhoi. Seoul, ROK: Gyeongin Munhwasa, 2006.

Sung, Chae-gi, Joo-hyun Park, Jae-ok Park, and O-bong Kwon. *bukhan gyeongjewigi 10nyeongwa gunbijeunggang neungryeok [North Korea's Decade of Economic Crisis and Capacity for Military Buildup]*. Seoul, ROK: Korea Institute for Defense Analyses Press, 2003.

Tellis, Ashley J. *Measuring National Power in the Postindustrial Age*. Santa Monica, CA: RAND/Arroyo Center, 2000.

Thakur, Ramesh. "New Zealand: The Security and Tyranny of Isolation." In *Strategic Cultures in the Asia-Pacific Region*, edited by Ken Booth and Russell B. Trood. New York, NY: St. Martin's Press, 1999.

Tsukamoto, Katsuichi. *kitachousengunto seiji [The North Korean Army and Politics]*. Tokyo, Japan: Hara Shobo, 2000.

US Department of the Army. *ADRP 3-0: Operations*. Washington, DC: US Department of the Army, October 2017.

———. *ATP 7-100.2: North Korean Tactics*. Washington, DC: US Department of the Army, July 2020.

US Joint Chiefs of Staff. *Joint Publication 3-0: Joint Operations*. Washington, DC: US Joint Chiefs of Staff, 17 January 2017.

Yi, Dae-keun. *bukhanguneun woae kudetareul haji ana [Why Don't the Korean People's Army Make a Coup]*. Paju, ROK: Hanul Academy, 2003.

Yoo, Yong-won, Beom-chul Shin, and Jin-a Kim, eds. *bukhangun sikeurit ripoteu [North Korea Military Secret Report]*. Seoul, ROK: Planet Media, 2013.

2 The strategic mindset

Since the formative years, the DPRK's strategic perceptions have been threat-based focusing on the US, ROK, Japan, and other allies. The DPRK describes the US as the antagonist and the natural enemy of Koreans for their imperialist occupation of South Korea that led to the nation's division and enslavement of South Korean citizens. The DPRK also targets the ROK and refuses to recognize them as a legitimate government for collaborating with the US and working against common "Korean" interests for peaceful unification. Japan is also a target, not simply for their former occupation of the Korean peninsula and the fact that Kim Il-sung's background stems from his anti-Japanese partisan campaign but also for Japan's hardline stance against the DPRK, close alliance relations with the US, and the massive US forces based there. But while the DPRK's perceptions toward the US, ROK, and Japan reveal what the regime targets, there are circumstantial variables that explain how the strategies were shaped and pursued. Thus, understanding the DPRK's strategy and strategic culture requires an assessment of how Pyongyang has viewed the internal and external developments through the lens of not only its strategy for unification known as the Three Revolutionary Forces for Reunification but also its sense of national identity.

The circumstantial developments

The establishment of the two Korean governments in 1948 both with the common goal of unification but with bipolar political identities and systems that mutually refused to recognize one another as legitimate governments made conflict inevitable. Kim Il-sung saw the situation as not only a threat but also an opportunity to become a national hero who expels the US forces and unifies the divided nation. By 1949, Kim Il-sung's preparations for the invasion of the South were well underway, working to strengthen the KPA's readiness while also collecting support from the socialist benefactors. But while the DPRK signed a economic and cultural cooperation agreement with the USSR in March 1949, Joseph Stalin was initially far from enthusiastic about Kim Il-sung's war plan.[1] Stalin finally agreed to the DPRK's invasion of the South in the spring of 1950 and promised to provide material support on the condition that the USSR is not directly involved in the war and that Kim Il-sung also gets Mao Zedong's agreement (to which Mao did).

Stalin's change of mind did not simply come from Kim Il-sung's persistence, but rather from the circumstantial developments, including the Chinese Communist Party's victory in China, the USSR's first nuclear weapons test, and also the view that the US was uninterested in defending South Korea.

Kim Il-sung was confident about victory for four reasons. First, the KPA had double the number of personnel and was much better armed than the ROK counterpart.[2] Second, Kim Il-sung and many other cadres believed that there were a significant number of leftists in the South who were waiting for the North Korean intervention to join the WPK cause. Third, Kim Il-sung felt he had sufficient backing from Beijing and Moscow not only for the invasion but also as guarantors of the regime. Fourth, Kim Il-sung assumed that the US would not intervene as it had withdrawn much of its forces from South Korea and drew the Acheson Line that outlined the US's defensive zone in East Asia but excluded the Korean peninsula.

On 25 June 1950, the DPRK launched its invasion against the South, calling it the "Great Fatherland Liberation War" – a supposed defensive crusade to save Korea from the US and Rhee Syngman to unify the Korean peninsula. Initially, the abrupt invasion was successful, swiftly taking over the key cities and cornering the ROK forces to the Busan perimeter. Fortunes reversed, however, with the amphibious landing by the United Nations (UN) forces at Incheon that divided the KPA and facilitated the rollback of North Korean forces back toward the Yalu River. The KPA most likely would have evaporated if it was not for the intervention by the Chinese People's Volunteer Army in October 1950 that pushed the UN forces back South. Following two years of fighting along the 38th parallel and tense negotiations, the two sides with the absence of the ROK (who sought unification) signed an armistice on 27 July 1953 that resulted in the confirmation of the division without solving its cause – at the cost of several million casualties and the separation of countless families.

Kim Il-sung's failed military campaign was the result of four major miscalculations. First, the US did intervene, leading the UN forces and a small Japanese maritime minesweeping unit to counter the DPRK. Second, the KPA's readiness was insufficient, particularly against UN air, naval, and amphibious operations. Third, the actual number of South Koreans who joined the North Korean cause was far below Kim Il-sung's expectations and many in fact fled or were evacuated to the South. Fourth, the DPRK struggled to reinforce the KPA or to fill the casualties, and those that joined were poorly disciplined and trained.[3] The miscalculations and failures were major lessons learned, forcing Kim Il-sung to reconfigure his strategy and become stricter about the conditions for unification under the WPK.

At the Central Committee of the WPK (CCWPK) Plenary Meeting in November 1954, Kim Il-sung argued that unification can only be achieved with a strong party and state government in North Korea and a leftist revolution in South Korea.[4] Then at the Eighth Plenary Meeting of the Fourth CCWPK in February 1964, the WPK adopted Kim Il-sung's Three Revolutionary Forces for Reunification that called for: strengthening revolutionary forces in North Korea, building revolutionary forces in South Korea, and strengthening the international revolutionary forces.[5] The strategy was much about the post-armistice developments such as Kim Il-sung's domestic political struggles, establishment of the US–ROK alliance, political-economic developments in South Korea, Japan's rearmament, the US and its allies' Cold War strategies, and uncertain developments among and within the socialist states. Under such circumstances, it was clear that communization and unification of the Korean peninsula simply through an armed invasion was implausible, and the goal could only be achieved if socialist forces in the South weaken the ROK government and rally the masses to join the DPRK cause, and if Pyongyang gains wide international support from fellow "revolutionary" states. On paper, Kim Il-sung's description of the Three Revolutionary Forces for Reunification was filled with arguments about what the DPRK must do or what must happen to achieve unification. At the same time, however, the Three Revolutionary Forces for Reunification revealed the DPRK's strategic perceptions and the factors Pyongyang is sensitive to, becoming a set of barometers that indicate not only its chances for unification but also the regime's survival.

The North Korean revolution

The so-called "North Korean revolution" is essentially about accomplishing the "Kim Il-sung revolution." But to understand the "Kim Il-sung revolution," one must start from the state's formative years. To this day, North Korean official accounts claim that Kim Il-sung solely led the liberation of the peninsula from Japan and the construction of the independent socialist state. Pyongyang's version of history, however, is tailored to legitimize Kim Il-sung's profile and the actual developments were far more complex.[6] In fact, Kim Il-sung did not return to the Korean peninsula until more than a month after liberation and did not make his first public appearance until the welcoming rally for the Soviet army held in Pyongyang on 14 October 1945. Kim Il-sung's long absence from the Korean peninsula was a disadvantage, making him a distant and largely unknown figure compared to some of the local political activists such as Hyon Jun-hyok, Jo Man-sik, Kim Gu, Lyuh Woon-hyung, and Pak Hon-yong.

Although the WPK quickly became the most effective and popular organization in North Korea, their establishment was far from straightforward. The WPK formed through the merger of various communist groups based on their geographical origins and associations: The Manchurian guerrilla faction led by Kim Il-sung; the Soviet faction consisting of USSR citizens of Korean ethnicity; the Yanan faction led by Kim Tu-bong, Mu Jong, and Pak Il-u who had close associations with the Chinese Communist Party and formed the New People's Party of Korea; the Kapsan faction made up of anti-Japanese partisans from the northern regions of the Korean peninsula; and the domestic Korean faction (otherwise known as the South Korean faction) led by Pak Hon-yong, who founded the Workers' Party of South Korea (WPSK) on 23 November 1946. The first merger took place on 10 October 1945 when the communists in northern Korea came together to form the Korean Communist Party Northern Korea Bureau (later became the North Korea Communist Party on 18 November 1945). Then on 28 August 1946, the North Korea Communist Party merged with the New People's Party of Korea to form the Workers' Party of North Korea (WPNK). When the WPNK and WPSK finally merged to become the WPK on 30 June 1949, Kim Il-sung was elected as Chairman, eclipsing the leading figures from rival factions, such as Ho Ga-i, Kim Tu-bong, and Pak Hon-yong. The various factions with diverse backgrounds and with little contact in the pre-liberation era were liabilities for the party's unity but were exploitable opportunities for Kim Il-sung to maneuver up the political ladder.

In the state, Kim Il-sung's power grew as *de facto* governance institutions were forming in North Korea. On 8 February 1946, an executive organ known as the Provisional People's Committee of North Korea was formed. Then in February 1947, the People's Assembly of North Korea held its first meeting, where the WPNK far outsized the Korean Democratic Party and the Chondoist Chongu Party. The *de facto* legislative body elected the members of People's Committee of North Korea on 22 February 1947 with Kim Il-sung as the Chairman and Kim Tu-bong as the Vice-Chairman. Kim Il-sung's positions in both the *de facto* executive and legislative branches enabled him to be appointed as the Prime Minister when the DPRK was officially established on 9 September 1948.

Several factors explain Kim Il-sung's rise. First and most significant was the USSR's endorsement of Kim Il-sung. Although Kim Il-sung was not part of the Soviet faction, his stint in the Soviet 88th Brigade was enough to gain the trust from the Soviet occupying forces. The relationship was based on convergent interests and mutual benefits. For the Soviets, although the 33-year-old Kim Il-sung lacked political experience and local credentials, he was a locally born Korean who was consistently cooperative and obedient, making him a "safe option" to implement Moscow's interests.[7] For Kim Il-sung, working with the Soviets was a vital but relatively burden-free option to gain political endorsement as the future leader of the country. Based on the convergent interests, the partnership was natural, leading to Stalin's selection of Kim Il-sung as the leader of North Korea.

Second, Kim Il-sung's political groundwork focused on establishing the armed forces while his rivals were preoccupied with political and bureaucratic affairs. For Kim Il-sung whose career was almost wholly in guerilla partisan activities, competing with experienced political activists and bureaucrats was no easy task and instead focused on establishing the military institution. Arguably, Kim Il-sung was the most effective in expressing and practicing *realpolitik* among the political figures in North Korea, matching with the post-liberation nationalistic atmosphere to win the hearts and minds of the public.

Third, economic developments and reforms during the early years also helped Kim Il-sung. Under the initial economic plans, North Korea not only took advantage of the various assets left behind by the Japanese and Soviet aid but also executed various reforms. The initiatives proved to be successful, with Pyongyang quickly industrializing and mobilizing the masses that allowed it to economically outperform its southern counterpart. While the DPRK's state construction was nascent, the contrasts with the economically troubled South and the perceived efforts to correct the unfairness of previous regimes helped Kim Il-sung gain further credibility as the leader of the new Korean state.

Despite the advantages, Kim Il-sung's position both in the state and in the party remained tenuous, and the failure to unify the Korean peninsula and near defeat in the Korean War had put his credibility in peril. To ensure his political survival, Kim Il-sung restructured the WPK during the Korean War through mass recruitment to dilute the influence of other factions.[8] Kim Il-sung also took preventative actions by blaming his domestic rivals for the disastrous military campaign. Immediately, Kim Il-sung targeted the South Korean faction of the WPK, purging figures such as Pak Hon-yong and Ri Sung-yop, but also Ho Ga-i from the Soviet faction for allegedly spying for the US and ROK. Senior military officers such as Mu Jong were also forced to take responsibility for the fall of Pyongyang in November 1950.

As Kim Il-sung tightened his authority, senior figures from the Soviet and Yanan factions became increasingly vocal in their opposition and attempted to oust the leader. In response, Kim Il-sung purged senior figures from the Soviet and Yanan factions in August 1956.[9] Despite Beijing and Moscow's efforts to calm Kim Il-sung, the WPK congresses in 1956 and 1961 resulted in the Manchurian faction dominating the party leadership. The Kapsan faction was initially spared from the political bloodbath given their general agreement with Kim Il-sung, but later in March 1967, Kim To-man, Pak Kum-chul, and Ri Hyo-sun were systematically removed when they challenged Kim Il-sung's growing personality cult and also proposed more economic-centered policies over Kim Il-sung's military-centric plans.[10] By the end of the 1960s, virtually all of Kim Il-sung's prominent rivals had been removed, further refining the CCWPK to those who pledged their unwavering loyalty to the leader.

Reconstructing the war-torn economy was also vital for Kim Il-sung to consolidate (and restore) his credibility. At the Sixth Plenary Meeting of the CCWPK on 5 August 1953, Kim Il-sung announced the Three-Stage Plan for Post-War Reconstruction – a nine-year scheme that aimed to rehabilitate and expand the DPRK's industrial capacity. Initially, the DPRK focused on consumer goods as prescribed by the Soviet and Yanan factions, but later in 1955, Kim Il-sung strong-armed the WPK to focus on the heavy industry sector.[11] Kim Il-sung also pressed forward with collectivized mobilization under the WPK through the Chollima Movement, which was essentially a carbon copy of China's Great Leap Forward, as well as the Chongsanri and Taean Work System that reshaped the agricultural and industrial sectors to be run by "technologically minded loyalist cadres."[12] While these measures were much about giving Kim Il-sung greater command and control over the state's economic development, there was a credible level of short-term success as seen in the rapid economic growth and industrialization.

Kim Il-sung also reformatted the DPRK's national identity by implementing an indigenous political software to iron out and deflect external and internal political influences that threatened the leadership. In the mid-1950s, the DPRK became louder in its call for the nation to defend itself against bureaucratism, dogmatism, flunkeyism, formalism, and revisionism. In 1955, Kim Il-sung gave speeches on the need for Juche (self-reliance), embarking on the steady refurbishment of the state's political, economic, and societal landscape.[13] In his lecture at the Ali Archam Academy of Social Sciences in Indonesia in April 1965, Kim Il-sung gave his first detailed explanation on the Juche idea by outlining its three pillars: Jaju Jeongchi (political independence), Jarib Gyeongje (economic self-sustenance), and Jawi Gukbang (self-defense).[14] Then at the First Session of the Fourth Supreme People's Assembly (SPA) in December 1967, Kim Il-sung promoted Juche's status from the ideology of the party to the "guiding principle" of the state.[15]

Draconian measures were also taken against the society, particularly with the Songbun political caste system introduced by the WPK in 1957. Broadly, Songbun divides the citizens into three categories and 51 subcategories depending on the family's background prior to liberation, activities during the formative years and the Korean War, connections with enemy states such as the US, Japan, and ROK, and also political loyalty and behavior. The class affects the individual and the family's education, occupation, rations, welfare, housing, and in worst cases could lead to imprisonment.

Although Kim Il-sung effectively removed the political fault lines that threatened his authority, the various developments in the 1950s and 1960s had outdated the original constitution. The original constitution adopted in 1948 was quite simple and mentioned little about the state's ideological nature, nor mandated a particular party, let alone an individual, to run the state. Thus, to constitutionalize the dictatorial Suryong system, the DPRK adopted the new Socialist Constitution at the First Session of the Fifth SPA in December 1972. The new constitution not only stated the WPK's exclusive rights to govern the state but also replaced the former cabinet-based system to one that is ruled by the Central People's Committee (CPC) commandeered by the President.

The fate of the Kim Il-sung-based system heavily depended on the succession process. For Kim Il-sung, the power successions in other communist states were far from inspiring. In the USSR, the troika of Lavrentiy Beria, Georgy Malenkov, and Vyacheslav Molotov that succeeded Stalin was soon ousted by Nikita Khrushchev who later unwound the Stalinist system. As for China, Beijing was rocked not only by Lin Biao's alleged plot to assassinate Mao Zedong in 1971 but also by the Gang of Four. Kim Il-sung knew well that mishandling the succession process would make him either a lame duck, or worse, erased from history. Given that Kim Il-sung was already beginning to be worshipped as a divine figure, the creation of a dynasty was seen as the only way to preserve the regime, selecting his first son Kim Jong-il as the heir to the throne. Although Kim Jong-il lacked both military and political experiences, he was endorsed by Kim Il-sung's closest aides such as Choe Hyon, Choe Yong-gon, Kim Il, and later O Jin-u. In 1973, Kim Jong-il was appointed as secretary in the CCWPK, and he headed the Three Revolutions Team Movement to strengthen his power base and was soon recognized as the successor within the CCWPK under the alias of "the Party Centre." Then at the Sixth WPK Congress in October 1980, Kim Jong-il assumed positions in the Standing Committee of the Political Bureau, Secretariat, and the Military Committee of the CCWPK (MCCCWPK), becoming the most senior party cadre after Kim Il-sung.

The implementation of the dynasty took politicization to new heights, particularly with the Three Revolutions Team Movement to bolster and enforce the Kim dynasty's legitimacy and personality cult. The DPRK constructed monuments and buildings including the Juche Tower, Arch of Triumph, Grand People's Study House, Rungrado May Day Stadium, as well as the bronze statues of Kim Il-sung. Given that the economy was already exhausted from the previous Seven Year Plan that centered on building the military-industrial complex and arms buildup, the measures to legitimize the dynasty system had detrimental economic effects. Both the Six Year Plan and the Second Seven Year Plan failed to fix the shortfalls of the economic plans to date, creating the trend where failures from preceding plans rolled over to the next. Pyongyang attempted to fill the shortfalls through the so-called "Speed Battle" campaigns that intensively mobilized and exhausted the masses to increase production, but with little meaningful effect. Even in attracting foreign capital, the introduction of the Joint Venture Law, establishment of Special Economic Zones, and other activities did little to solve Pyongyang's mounting foreign debt.

The transition to the second-generation leadership accelerated into its final stages in the early 1990s with Kim Jong-il becoming the SCAF at the Nineteenth Plenary Session of the Sixth CCWPK in December 1991, and the Chairman of the NDC at the Fifth Session of the Ninth SPA in April 1993. In hindsight, Kim Jong-il's promotion was timely for Pyongyang given the death of Kim Il-sung on 8 July 1994. Yet the second-generation leadership did not come into immediate effect, with Kim Jong-il calling for a three-year mourning period where almost all major party and state activities were put on hiatus, presumably to plan strategies and make realignments with minimal risks. Like his father, Kim Jong-il relied on politicization to bolster his authority by unveiling Songun Jeongchi (military-first politics). According to the DPRK, Kim Jong-il first introduced Songun during his visit to a KPA unit in January 1995 where he described the military's pivotal role to protect socialism and "complete" the Juche-ist revolution. In the following years, Songun was implemented as the means of ensuring the DPRK's survival in the "arduous march" era.[16] Still, Songun was not simply about empowering the military *per se*, but rather as Yi Dae-keun aptly describes, politicized the military while militarizing the society.[17] Thus, Songun was more of a means for Kim Jong-il to bolster his leadership by regimenting the society while ensuring the military's loyalty.

The economy continued to stall in the 1990s. The Third Seven Year Plan was shelved after a three-year extension, and the socio-economic privations were becoming more serious and widespread. Moreover, the DPRK poorly responded to the natural disasters in the mid-1990s, leading to the collapse of the agricultural sector and the eventual suspension of the Public Distribution System. Citizens in the less-urbanized areas which account for approximately 85% of the total population suffered the most because of poor access to critical infrastructures and resources, exposing them to severe poverty, malnutrition, and starvation. In response, the citizens created illicit capitalist markets that began in the form of small vendors selling daily goods, but with some expanding to larger stores, factories, real estate, and transportation. While the private markets did not empower the citizens, new norms were undermining the state system. First, citizens were now less dependent on the state-controlled economy. Second, in the economic context, the Songbun caste system became less relevant as the private markets were about entrepreneurial skills rather than class. Third, corruption became a norm, where entrepreneurs bribed officials to escape punishment or to make deals, while the officials depended on the kickbacks to fill their meager income. Facing this new reality, the government allowed a certain level of market activities while preserving state control through the July 1 Economic Management Measures in 2002 and the currency redenomination in 2009, albeit both of them failed, further highlighting the government's incompetence.

Speculations about the successor to Kim Jong-il emerged in 2008 due to his ailing health and absence from the 60th anniversary of the state's founding. Behind the scenes, Kim Jong-il's third son Kim Jong-un was being prepped as the successor, while songs and other propaganda that subtly declared the new era under the leadership of a young general began to emerge. On 27 September 2010, Kim Jong-il issued Order Number 0051 that promoted 40 KPA officers including Kim Jong-un to the rank of General. Then on the following day at the WPK Conference, Kim Jong-un was appointed as the Deputy Chairman of the CMCWPK. Much like Kim Jong-il in the 1980s, Kim Jong-un began to publicly accompany his father in various guidance tours, promoting his presence as the upcoming leader. Immediately after Kim Jong-il's death on 17 December 2011, Kim Jong-un was appointed as the SCAF and *de facto* leader of the party and the state. Yet unlike the previous transition, the mourning period was only 11 days, partly for the scheduled celebration of the centennial of Kim Il-sung's birth in April 2012, but more importantly, because the three years of inactivity after Kim Il-sung's death exacerbated the problems in North Korea. Kim Jong-un's leadership was officiated in April 2012, when he was appointed as the First Secretary of the WPK at the Fourth WPK Conference and also as the First Chairman of the NDC at the Fifth Session of the Twelfth SPA.

Initially, there were questions about Kim Jong-un's ability to lead, not least because of his young age but due to the fact that his official ascension spanned only a year and three months (significantly shorter than the two decades from Kim Il-sung to Kim Jong-il), and also that the young leader was only the Deputy Chairman of the CMCWPK without any official positions in the NDC or the Presidium of the Political Bureau of the CCWPK at the time of Kim Jong-il's death. Expectedly, Kim Jong-un depended on politicization to legitimize his leadership. At the Fourth Conference of the WPK in April 2012, Kim Jong-un introduced "KimIlsung-KimJongilism," which essentially merged the Juche and Songun ideologies to demonstrate how he will continue the legacies of the predecessors. At the same time, Kim Jong-un has proactively relieved, retired, or banished a high number of senior party, state, and military cadres with the most notable being the publicly announced execution of Kim Jong-un's uncle Jang Song-thaek (husband of Kim Jong-il's younger sister Kim Kyung-hui) in December 2013. The personnel changes reflected not only Kim Jong-un's unforgiving style of leadership but also his efforts to find the right combination of cadres. Moreover, there has been greater clarity in governance under Kim Jong-un, with party activities becoming more formalized, and the replacement of the NDC with the SAC in 2016. Kim Jong-un also promoted his WPK leadership profile to demonstrate his growing authority, becoming the Chairman at the Seventh WPK Congress in May 2016 and then the General Secretary at the Eighth WPK Congress in January 2021.

While Kim Jong-un has inherited the centralized and politicized leadership practices of his predecessors, he has also pursued performance-based legitimacy. One key policy of the Kim Jong-un regime has been the Byungjin (parallel) line unveiled in 2013, promising the parallel development of nuclear weapons capabilities and the economy. For Kim Jong-un, economic development would make or break his leadership, as genuine improvements will legitimize him as a successful leader that modernized the DPRK while failing to do so would undermine his credibility and exacerbate the failures inherited from his predecessors. Indeed, Kim Jong-un has taken some steps toward economic improvement and has, in fact, encouraged market activities by introducing new commercial laws. Nevertheless, the measures to date have been about modernizing the economy under the government's control without reform. Moreover, the economic developments since the 2010s have been far from smooth. In January 2011, Pyongyang announced its first economic plan in 18 years – Ten-Year State Strategy Plan for Economic Development – that was then replaced by the Five-Year Economic Development Strategy announced at the Seventh WPK Congress in May 2016 that fell far short of its target. The troubles do not simply owe to poor management but also to international sanctions, natural disasters, and COVID-19. Much remains uncertain about the fate of the new five-year economic plan unveiled at the Eighth WPK Congress in January 2021, although the regime's policies and current circumstances suggest that the obstacles are high.

Despite being portrayed as a tough, hands-on leader, Kim Jong-un's dubious state of health and his occasional long absences from the public eye have raised speculations about his leadership. While Kim Jong-un remains to be well in charge at the time of writing, his younger sister Kim Yo-jong has gained much attention for her role as the right hand of the leader and also in inter-Korean relations – both in dialogues and verbal attacks against the ROK. Much is unknown as to Kim Yo-jong's exact position, although she served as an alternate member of the Political Bureau of the CCWPK between 2017 and 2021, and is known to have served in various positions within the key party organs. Still, while Kim Yo-jong may not have an official title in the highest echelons of the regime, it is widely believed she still plays a key role in the leadership. Kim Yo-jong's future succession would make sense in the "bloodline" context, and much would pivot on the positions she will gain in the party, state, and the military in the coming years.

Over the decades, the three generations of the Kim dynasty called for the party and the state's unity to achieve the "socialist revolution." But the key characteristic of the so-called revolution was the transformation of the state and the society into an absolute top–down system based on the dynasty. Much is seen in how the regime describes itself in what Armstrong termed as "familial" and "organicist imagery."[18] The country is metaphorically described as a living being, with the leader as the brain, the WPK as the nerves, and the society as the cells. The hierarchy is also expressed as a family, with Kim Il-sung being the father, the WPK being the mother, and the citizens as the children. The revolution was far less about the communist revolution on the Korean peninsula, but about erasing the leaders' enemies while also transforming the system into a totalitarian dictatorship. Consequently, the Kim dynasty's revolution narrows the definition of unification, where the only form allowed is the one that guarantees the regime's authority.

The South Korean revolution

The "revolutionary forces in South Korea" principle is about the awakening and organization of radical leftist forces in South Korea that influence the masses to side with the WPK. The key idea was that unification (or even coexistence) under WPK terms would be better achieved if the South Koreans develop their own "revolutionary forces." Yet although the principle sounds passive, the DPRK has historically sought to exploit the political developments in the South while also agitating the South Korean leftists through a variety of counterstability operations.

For North Korea, South Korea is not simply the other half of the state but also a destination – explaining why the DPRK claimed Seoul as its official capital until the constitutional amendment in 1972. Since the very early years, South Korea was the center for much of the political activities on the Korean peninsula. In November 1946, the New People's Party, the People's Party of Korea, and former members of the Communist Party of Korea came together to form the WPSK – the party that purported to merge with the WPNK under a unified Korean state. The WPSK led various uprisings in South Korea but were suppressed and outlawed from taking part in the Constitutional Assembly elections held in May 1948. Given that the right-wing forces were certain to come out victorious in the more populated South, both the WPNK and the WPSK viewed there was much to lose and nothing to gain from the UN-sanctioned elections. Moreover, although Pak Hon-yong had convinced Kim Il-sung that there would be a mass exodus of South Korean leftists that would join the DPRK, the actual chances were slim, particularly with the series of massacres in South Korea against anyone suspected of being leftist.

Rhee Syngman became ever more authoritarian after the armistice and called for unification under the slogan of Bukjin Tongil (northward march for unification). That said, South Korea was economically battered, and Rhee had lost significant amount of credibility and was scraping to survive through bribes and rigged elections. To exploit the situation, the DPRK indirectly sent proposals to the South for steps toward unification, calling for the withdrawal of all foreign armed forces, arms control, as well as some grassroots initiatives such as communication between divided families, and even a joint team at the 1960 Rome Olympiad.[19] At the same time, however, Kim Il-sung argued at the Plenary Session of the CCWPK in November 1954 for the need to rebuild and strengthen the KPA. Thus, Kim Il-sung's proposals for arms control and rapprochement were not about peace, but rather about capping the military developments in the ROK so that the DPRK can maintain the upper hand.[20]

The developments in South Korea continued to go in Kim Il-sung's favor. By 1960, student-led anti-government rallies were widespread, climaxing with the April Revolution that led to Rhee Syngman's resignation. Chaos continued under the parliamentary Second Republic led by Yun Bo-seon and Chang Myon that failed to effectively govern and stabilize South Korea. Although Pyongyang did not intervene, Kim Il-sung continued to indirectly agitate the political left in South Korea by propagating Pyongyang's political-economic success under the WPK and proposed the confederate model of unification.[21] It is unknown whether Kim Il-sung refrained from intervening out of fear of suffering another military failure, lack of support from China and the USSR, or just simple miscalculation. Nevertheless, the DPRK had missed a prime opportunity, only sitting back to watch the Second Republic fall to the coup d'état led by Park Chung-hee on 16 May 1961.

Park Chung-hee established a government that was staunchly anti-communist and practiced authoritarian developmentalism to strengthen the ROK's state power. In particular, Park Chung-hee's focus on industrialization propelled the ROK economy into exponential growth that soon surpassed that of the DPRK. Park Chung-hee was also geostrategically proactive, outreaching to Japan, and also deploying troops to the Vietnam War. In response, Kim Il-sung at the WPK Conference in October 1966 called on the "revolutionary forces" in South Korea to challenge the Park Chung-hee regime, reflecting the level of alarm he had toward the developments. Given the disadvantages in facing a growingly stronger ROK, the DPRK carried out asymmetric attacks against the South, including the Blue House raid in January 1968 in an attempt to assassinate Park Chung-hee, the Uljin-Samcheok landings in October 1968 to agitate and recruit the political left in South Korea, and the hijacking of a South Korean airliner by a North Korean agent in December 1969. Moreover, the KPA seized a US Navy vessel in January 1968 and also shot down an EC-121 in April 1969, raising regional tensions. The chain of attacks by Pyongyang further emboldened Seoul's hardline strategies, including plans for its own black operations to assassinate Kim Il-sung with the "684 Unit" that was later aborted.

Dramatic changes took place in the early 1970s, with the international mood for détente and the DPRK's quest to secure cash opening a period of direct contact between the two Koreas. Interactions began with inter-Korean Red Cross meetings (that were attended by political and security officials from both sides) that then led to working-level talks and mutual visits by senior officials.[22] While the two Koreas did not hold a summit between the leaders, they issued the Joint Communique on 4 July 1972 that pledged their efforts for peaceful unification based on national unity free from foreign involvement. In practice, the initiatives were much about opening up direct communications between the two Koreas as opposed to taking substantive steps toward unification. Still, the two Koreas remained starkly divided, not simply due to the lack of trust but also because Kim Il-sung and Park Chung-hee were competing to bolster their respective authoritarian rule. Indeed, there were some positive developments for Pyongyang, particularly when the US was mulling the withdrawal of its troops from South Korea. Yet this only triggered Park Chung-hee to pursue more self-reliant means to defend itself against the North, even going to the point of initiating the clandestine, but later aborted nuclear weapons program. There were also conflicts between the two Koreas. On 15 August 1974, Mun Se-gwang, a pro-DPRK Korean from Japan attempted to assassinate Park Chung-hee, resulting in the death of Yuk Young-soo (the first lady). Moreover, there were military conflicts, with the ROK discovering a number of the DPRK's infiltration tunnels along the demilitarized zone (DMZ) as well as the axe murder incident at Panmunjom on 18 August 1976. Thus, the inter-Korean relations soon slipped back into conflict and talks at all levels had stopped.

The assassination of Park Chung-hee by his key aide on 26 October 1979 started a new period of political turmoil. Although Choi Kyu-hah was technically the president, the actual power was held by Chun Doo-hwan who staged an internal coup on 12 December 1979 to bolster his positions in both the military and the intelligence organs. Pro-democracy movements in South Korea were growing with greater intensity but were ruthlessly suppressed, most notably with the Gwangju Uprising in May 1980. Chun Doo-hwan used the momentum of the coup and the suppression campaigns to establish and preside the Fifth Republic that came into effect in March 1981. Chun Doo-hwan not only strengthened his rule but also eased restrictions on recreation to distract the public from politics. But by this time, the credibility and effectiveness of authoritarian developmentalism was corroding and South Korea had to democratize. The South Korean economy was booming, and Seoul was set to host the Summer Olympiad in 1988, reflecting the ROK's international attention and status. In June 1987, nationwide pro-democracy protests took place, leading to the establishment of the democratic Sixth Republic.

The 1980s was marked by the DPRK's counterstability operations to delegitimize the ROK government and to taint their international status. The DPRK conducted a number of terrorist attacks against the ROK including the Rangoon bombing in an attempt to assassinate Chun Doo-hwan on 9 October 1983 and the bombing of Korean Air Flight 858 on 29 November 1987. The DPRK also agitated the far-left in South Korea who did not hide their admiration of the Juche idea and chanted a renewed sense of ethnocentric nationalism that criticized the ROK government to date for their "pro-US" and "pro-Japanese" principles. For the DPRK, engagement with the far-left groups was not so much about instigating an immediate revolution, but rather a political investment to create advantageous circumstances in the future. Despite the tensions, however, there were moments of inter-Korean interactions during the 1980s, with a number of working-level dialogues and some grassroots exchanges including the reunion of separated families. Still, the dialogues and exchanges did not lead to any substantive steps toward peace and reconciliation.

Governments under the democratic Six Republic presented both opportunities and challenges for the DPRK. Roh Tae-woo was diplomatically pragmatic, establishing relations with the USSR in 1990 and with China in 1992 under the auspices of Nordpolitik. The aim was not just about diversifying Seoul's diplomatic channels but also to contain and lure Pyongyang into dialogues. Roh Tae-woo also took steps in inter-Korean relations, proposing the formation of a Korean Commonwealth and lifted restrictions on inter-Korean trade. Clear steps toward conflict prevention and reconciliation were seen in the early 1990s with the signing of the Agreement on Reconciliation, Non-aggression and Exchanges and Cooperation between the South and the North in 1991, and the Joint Declaration of South and North Korea on the Denuclearization of the Korean Peninsula in 1992. Although problems remained, the two Koreas had taken steps to turn the relations from one of confrontation to one of coexistence.

Much changed, however, during the Kim Young-sam administration. In March 1993, the DPRK threatened to withdraw from the Non-Proliferation Treaty after the International Atomic Energy Agency (IAEA) demanded the inspection of two nuclear waste storage sites. While Kim Young-sam stood defiant against the DPRK's behavior, he also offered to hold a summit with Kim Il-sung. The summit did not materialize due to Kim Il-sung's death in July 1994, and Pyongyang renewed its bellicose attitude against Kim Young-sam for not sending condolences to the deceased leader. Even though the US and the DPRK signed the Agreed Framework in October 1994, tensions between the two Koreas worsened, most notably in September 1996 with the infiltration by a DPRK submarine on the eastern coast of South Korea.

New developments came under the liberal Kim Dae-jung administration that prioritized engagement and cooperation to change the DPRK's attitude. Pyongyang positively responded to Kim Dae-jung's policies, leading to the first inter-Korean summit in 2000 that kick-started various cultural and economic initiatives. The rapprochement policy was inherited by the more progressive Roh Moo-hyun who proactively increased the amount of aid to the DPRK and also called for Jaju (independence) – an idea that almost mirrored Juche. Roh Moo-hyun's approach was clearly divergent from those of the George W. Bush administration who took a hawkish stance against the DPRK, and the strategic disconnects in the US–ROK alliance became evident, further straining the relationship already undermined by the tragic killing of two South Korean schoolgirls by US military personnel in June 2002.

The democratization and developments in the ROK's state power, combined with the policies pursued by both Kim Dae-jung and Roh Moo-hyun significantly influenced South Korean perceptions. While some remnants of anti-communist policies and perceptions remained, much of the demonizing caricatures and taboos relating to the DPRK began to fade, and perceptions toward the North and unification were diversifying. Many South Koreans saw (and still see) the North as the primary threat, but in ways different from the past, and the scenario where the small, impoverished country would invade and kill fellow Koreans in the much more powerful South was beginning to be perceived as unlikely. The DPRK's bellicose actions and military provocations also began to resonate less, and an increasing number of South Koreans began to view them as routine attention-seeking that posed limited direct harm. As for unification, while the sense of ethnic bond remains strong, many South Koreans also feel the difference with the North, seeing unification as a long-term process and supporting engagement as means of conflict prevention and reconciliation. At the same time, the far-left who were politically and socially censored in the past became more vocal and influential, with some starting to make their way into politics. Even though far-left ideas were subject to much stigma in the past, such perceptions began to be seen by some as refreshing alternatives to the old-style conservative politics.

The offerings and loose conditions promised by the Kim Dae-jung and Roh Moo-hyun administrations combined with the greater flexibility in South Korean public opinion served as opportunities for the DPRK to make gains with minimal risks. Of course, the DPRK had not let their guard down, wary of the fact that the ROK was far superior in state power and therefore had the potential to exercise significant amount of leverage. Nevertheless, the DPRK exploited the liberal-progressive ROK administrations. Days before Kim Dae-jung was set to visit Pyongyang in June 2000, the DPRK abruptly canceled the inter-Korean summit, and only agreed to go ahead after Seoul reportedly paid approximately USD 500 million to Pyongyang. The DPRK also pushed the envelope in security issues, unilaterally declaring a new border in the Yellow Sea contravening the Northern Limit Line (NLL). Military conflicts also took place, with two naval clashes in the Yellow Sea in June 1999 and June 2002. Moreover, the DPRK conducted its first nuclear test on 9 October 2006, clearly dishonoring the 1992 inter-Korean agreement. Despite the hostile acts, both Kim Dae-jung and Roh Moo-hyun continued to engage with Pyongyang, to which the DPRK affirmed that the liberal-progressive ROK governments present plenty of opportunities for exploitation.

Inter-Korean relations deteriorated rapidly after the inauguration of the conservative Lee Myung-bak administration in February 2008. Lee Myung-bak vowed to engage with Pyongyang on the condition of the North's denuclearization, economic reform, and economic growth. Angered by the conditional approach and Lee Myung-bak's efforts to strengthen the US–ROK alliance, the DPRK heightened its brinkmanship diplomacy, conducting nuclear and missile tests in 2009. Yet as it became clear that the ROK was not going to change its hardline stance, the DPRK resorted to actual assaults. On 26 March 2010, the DPRK sank the ROK Navy corvette Cheonan in the Yellow Sea killing 46 sailors on board. Then on 23 November 2010, the KPA shelled Yeonpyeong Island that killed two ROK marines and two civilians. Tensions continued to run high during the Park Geun-hye administration, with North Korean weapons tests, suspension of the Kaesong Industrial Complex in April 2013, and a landmine explosion on the border in August 2015 that maimed two ROK soldiers.

The impeachment of Park Geun-hye in 2017 led to the election of the progressive Moon Jae-in administration which included some of the far-left activists from the 1980s. The conservatives' loss of credibility meant that the progressives will remain strong for the time being, as proven in the National Assembly election in April 2020. For the DPRK, the political circumstances in the South showed greater opportunities to push the envelope. Initially, Pyongyang conducted a flurry of missile tests in 2017 but in the following year switched to charm offensive mode, leading to the DPRK's participation in the Pyeongchang Winter Olympiad and the three inter-Korean summits that vowed to end the Korean War and take steps toward unification. The two Koreas also inked a military agreement in September 2018 that purported to prevent conflict but was in essence detailed addendums to the armistice. But then in 2019, Pyongyang switched back to bellicose mode, blasting a torrent of insults toward the Moon administration, demolishing the inter-Korean liaison complex in Kaesong (which the ROK paid USD 15 million for the construction and renovation) on 16 June 2020, as well as murdering and burning the body of a ROK citizen in September 2020 who had entered North Korea. Despite the DPRK's hostile stance and domestic criticisms for the failures in the policies vis-à-vis Pyongyang, the Moon administration maintained its policies of engagement and appeasement.

While significant enhancements are seen in the ROK Armed Forces' readiness in recent years, the liberal-progressive ROK administrations have responded to the DPRK's bellicose attitudes in softer ways while also avoiding issues regarding Pyongyang's human rights abuses or dictatorial politics. The rationales of the Moon administration and the ruling party are mixed. While there are some who pursue close relations with the DPRK for ideological reasons, much is also about domestic politics, fearing that anything that jeopardizes dialogues or provokes harsh responses from Pyongyang would only prove the failure of the administration's engagement and appeasement strategies. Yet much have ended with substantive gains for the DPRK and symbolic gains for the ROK. The DPRK not only refused to correct their bellicose behavior but also has sadistically exploited the liberal-progressive ROK administrations by continuing to push the envelope, forcing Seoul to bend the norms and policies to meet their demands, help set up DPRK–US dialogues, but then turning their backs when the job is done or reverting to threats when they are dissatisfied. Naturally, such circumstances have created divisions within South Korea. On the one hand, critics have claimed that the ROK has compromised its security interests and the US–ROK alliance for the sake of dialogues that were later exploited by the DPRK. On the other hand, proponents argue that conflict has eased and that one day the engagement and appeasement will come to fruition. For the DPRK, a troubled ROK government (particularly a liberal-progressive administration) is pleasing, as this provides more exploitable opportunities.

Despite the various opportunities, the risks for intervention were too high for Pyongyang. Above all, even under the liberal-progressive governments, Seoul had always remained cautious of Pyongyang and the US–ROK alliance remained intact. But the other major factor was about politics, where the DPRK viewed that the "revolutionary forces" in South Korea had never matured enough for Pyongyang to intervene. Even during the times of authoritarian regimes, political opposition forces had always existed in South Korea, taking part in both presidential and general elections (albeit severely disadvantaged). Meanwhile in the North, Pyongyang continued to bolster the dictatorial system under the Kim dynasty by erasing any opponents. As time progressed, the differences between the two Koreas became increasingly apparent, particularly with the democratization of the ROK in 1987. To be sure, the DPRK was not pushing for democratization in South Korea, but rather saw democratization as a means that would provide opportunities for the far-left. But while the far-left have certainly gained greater influence in recent decades, the South Korean society remains to be diverse, with a rich mix of conservatives, progressives, as well as centralists. The political developments in the two Koreas could be no more divergent, making the DPRK increasingly unable to win the narrative over the ROK as the better political option, severely undermining the plausibility of intervention, let alone full unification under the WPK.

The international revolution

The "strengthen the international revolution" concept is about not just being part of the global socialist revolution but also collecting recognition and support for the DPRK cause. Much was shaped by the DPRK's view of the post-1945 geopolitical situation. Although Japan had now left the Korean peninsula and was no longer a credible threat (at least for now), the new situation in East Asia was a dire threat, epitomized by the anti-communist developments in South Korea, US presence in the western Pacific, and the Chinese Civil War that could potentially lead to the anti-communist Kuomintang to rule China. In the worst case, the DPRK could be wedged by the two anti-communist states that would significantly undermine its survival. Indeed, the communists prevailed in China, and socialism was spreading in various parts of Asia that relieved the DPRK's concerns. Yet at the same time, Pyongyang was well aware that its fate would pivot on whether it can gain international recognition and support.

The DPRK's diplomatic network in the initial years focused almost wholly on communist states. Other than the USSR and China, the DPRK established diplomatic relations with a number of states before the Korean War, including Albania, Bulgaria, Czechoslovakia, East Germany, Hungary, Mongolia, Poland, Romania, Vietnam, and Yugoslavia.[23] Thus, although the DPRK was not part of the UN-sanctioned elections and fought against the UN forces in the Korean War, it was able to get sufficient support from fellow socialist states that were growing in numbers in the post-1945 era. The DPRK was most assured when there was some level of strategic cohesion between China and the USSR – particularly leading up to the Korean War. Yet the Beijing–Moscow–Pyongyang triad was short-lived, and the DPRK soon became concerned about the developments within, and between China and the USSR. Soon after the death of Stalin in 1953, Nikita Khrushchev sought to unwind Stalinism and even pursued peaceful coexistence with the US, leading to significant chilling of relations between China and the USSR. China not only upped the voltage of its nationalistic and revolutionary ideals, but also openly and combatively criticized the USSR, to which Moscow responded by withdrawing its military and industrial assistance to Beijing.

The DPRK had become cautious about the influences of the political changes in the USSR, particularly with the factional incident in 1956 that nearly led to the ousting of Kim Il-sung. Nevertheless, the Sino–Soviet conflict created major dilemmas for the DPRK as Pyongyang could neither afford to be embroiled in the Sino–Soviet conflict nor become abandoned by the two benefactors. To ensure continued support, the DPRK signed alliance treaties with both China and the USSR in July 1961, with the DPRK–USSR Agreement on Friendship, Cooperation, and Mutual Assistance, and the China–DPRK Mutual Aid and Cooperation Friendship Treaty. At one point, the DPRK even tried to portray itself as a mediator and called for unity among all socialist states.[24] Yet the DPRK was not always neutral in the Sino–Soviet split, and periodically switched to lean toward the power that provided more benefits and presented less risks.

Initially, the DPRK distanced itself from the USSR and sided with China. In part, it was about politics, with concerns about Khrushchev's de-Stalinization possibly seeping into the DPRK and undermining Kim Il-sung's rule. Yet the bigger factor was the questionable credibility of the USSR as an ally. The aforementioned security agreements signed with both China and the USSR in 1961 were different in nature, where the pact with Beijing was set for an indefinite period, while the one with Moscow was only valid for ten years and then renewable every five years upon agreement. Pyongyang, therefore, viewed that the Soviets were less reliable, and affirmed its skepticism after witnessing the USSR's weak actions in the Cuban Missile Crisis of 1962. Although the USSR did not respond as harshly against the DPRK as it did against China, Moscow temporarily suspended its aid to Pyongyang.

The honeymoon between China and the DPRK was short-lived, leading to a revival in DPRK–Soviet relations after Alexei Kosygin's visit to Pyongyang in February 1965. One of the factors was the Cultural Revolution in China, where the DPRK was offended by the Red Guards who vocally labeled Kim Il-sung as a revisionist. Yet the larger factor was about security, where Beijing not only was inward-looking but also had limited capacity, making them far from generous in providing aid and also questionable in the actual benefits. The downturn in relations was evident with China's absence from the Military Armistice Commission from 1966 to 1971, to which Tsukamoto describes as a "barometer" to measure the state of Beijing–Pyongyang relations.[25] For the DPRK, the souring of relations with China raised concerns, but the thawing of relations with the USSR meant that the previously suspended aid from Moscow will resume, allowing further access to critical technologies for not only the heavy industry sector, but also the nuclear and missile projects that were vital to modernize the KPA.

By the end of the 1960s, the DPRK's relations with the USSR re-deteriorated, thereby shifting back toward China. Beijing rewarded Pyongyang's move by providing lucrative military aid and other provisions including an oil pipeline constructed in 1976. But more importantly, China was now growing to be a world power, gaining recognition in the UN as "the China" in 1971 and taking a seat in the UN Permanent Security Council. China's global status was a double-edged sword for the DPRK. On a positive note, China's growing leverage and status in international affairs gave greater assurances for the DPRK. Yet at the same time, China was establishing relations with a cohort of countries including the DPRK's enemies – namely Japan in 1972 and the US in 1979. Moreover, the China–Vietnam border conflict in 1979 instilled a sense of fear in the DPRK that China could turn against them for the sake of its geopolitical and geoeconomic interests. Hence, the relations during this time was far from cozy, and the relations soured again by the 1980s.

Despite shifting back to the USSR, there were both positives and negatives for the DPRK. In positive terms, the renewed conflict between the USSR and the US provided the DPRK with some degree of faith in the Soviets. The USSR provided much need aid to the DPRK and in return, the Soviets were granted access to North Korean airspace and ports for their operations in the Far East.[26] Yet by this time, the USSR's state power was steadily contracting, not only disadvantaging it in the balance against the US but was also struggling to keep the Union intact. Major changes were taking place under Mikhail Gorbachev, most notably with Glasnost and Perestroika that not only reformed the USSR but also contributed to its eventual dissolution on 26 December 1991. The collapse of the USSR was detrimental for the DPRK as it meant not only the loss of an economic and security safety net but also the opening of the post-Cold War era that would bring greater geopolitical challenges.

The DPRK's maneuvers in the Sino–Soviet split can be explained by three factors. First, the DPRK was always cautious about the contagious effects of the political developments in China and the USSR, steering away from the country that is going through some kind of domestic political trouble. Second, the DPRK distanced itself from the power that showed questionable credibility as an ally. Third, much was also to avoid patronizing pressure by one state to confront the other. Despite the DPRK's calculative maneuvers, it had never completely burned bridges with neither of its benefactors. After all, the DPRK not only needed both China and the USSR for material gains but also as diplomatic and strategic buffers against the US and its allies. Moreover, the DPRK never looked at the two allies in the same way, where on the one hand, China seemed to be a more politically devout partner and with the "blood relations" from the Korean War, while, on the other hand, the USSR offered more modern technologies and resources. Hence, the DPRK's oscillation between the two allies demonstrated how Pyongyang always sought to gain benefits from both benefactors while also trying to keep some distance to deflect any ramifications.

Both Beijing and Moscow were indeed frustrated by Pyongyang's opportunistic maneuvers and its militarist adventurism toward the US, ROK, and Japan. Nevertheless, the two benefactors never abandoned Pyongyang as doing so would bring little gains and only further complicate their geopolitical positions. Yet as time progressed, geopolitics and geoeconomics were beginning to work under new norms that were no longer bound by Cold War norms, with both China and the USSR viewing aid to the DPRK as "politically unnecessary and economically unjustifiable."[27] Making matters worse for the DPRK, by the end of the Cold War, both China and the USSR were working to normalize ties with the ROK. Pyongyang not only saw Beijing and Moscow's maneuvers as betrayal but also faced the reality that it was diplomatically outmaneuvered by Seoul. Against this backdrop, the chances of Beijing and Moscow supporting Pyongyang's forced unification of the Korean peninsula had diminished significantly.

Although the relations with China and the USSR were the main priorities during the Cold War, Pyongyang also actively outreached to, and even tried to mentor newly independent states, arguing that young states should develop through self-reliance to survive in the age of great power competition.[28] During the 1960s and 1970s, the DPRK established significant number of diplomatic relations with states in Southeast Asia, Middle East, and Africa. In particular, the DPRK saw the Non-Aligned Movement as a good platform to collect recognition and support. In some cases, the relations even extended into military partnerships, aiding the Communists in the Vietnam War, the Arab states in the Yom Kippur War, and government forces in the Ethiopian Civil War. Of course, there were strategic agendas such as gaining resources, as well as quantitatively collecting diplomatic recognition over the South. Initially, Pyongyang's maneuvers were successful, as quite a number of the young states admired the DPRK's ability to confront the US while establishing and maintaining its own independent identity. Yet as these states began to focus more on development, they started to pursue relations with states that offered substantive benefits or those that would serve as better models of development (namely Japan and the ROK) as opposed to the DPRK, which was proving to be a failed economic state that still practiced outdated and nepotistic politics. Even regarding the Non-Aligned Movement, the DPRK's strategic ambitions and bellicose tactics contradicted the group's more moderate and pragmatic vision.[29]

The end of the Cold War meant a drop in the number of states that ideologically and materially supported the DPRK. The actual political impact of the fall of socialist regimes on the DPRK (and for Asian communist states with the exception of Mongolia) was limited. After all, the DPRK had built a system that was far more centralized and politicized and never integrated into the socialist network like the East European states. But besides the critical loss of support, the major problem was that the post-Cold War order was working in ways that made the DPRK more vulnerable. First, globalization and economic interdependence became new norms, and East Asia was becoming a key region in the international economy. Second, the combined effect of economic growth and regional security uncertainties triggered the reconfiguration of defense policies and capabilities in China, Japan, ROK, and Taiwan, consequently upping the level of strategic competition in the region. Third, the ROK was now much more powerful and integrated into the global community than the DPRK, further isolating the Stalinist state.

The degree of complexity and fluidity of the post-Cold War environment was far beyond the DPRK's capacity, forcing Pyongyang to recalibrate and even compromise its diplomatic strategies. The DPRK agreed (reluctantly) to join the UN along with the ROK as separate states in September 1991. In contrast to the Cold War when the DPRK worked to internationally gain legitimacy over the South, in the post-Cold War era Pyongyang was fighting to stay relevant in the international community, even outreaching to the western and capitalist states. The DPRK even renewed its efforts to normalize relations with Japan, although nothing materialized due to not only Pyongyang's continued military provocations and the development of WMD, but also Kim Jong-il's admission to the abductions of Japanese nationals. In total, the DPRK established (or in some cases reestablished) diplomatic relations with nearly 60 countries since the 1990s.[30] Yet the increase in diplomatic relations and membership to the UN did not lead to the DPRK's integration into the international community as Pyongyang also remained isolated to buffer itself from unwanted external influence. Moreover, the states that established diplomatic relations with the DPRK did so in part to diplomatically contribute to solving the crisis and for possible trade benefits in the future, or simply because there were no reasons to not do so, as opposed to supporting Pyongyang's cause.

The DPRK also tried to gain legitimacy and support through public diplomacy, seen with foreign language publications, interactions with foreign media, hosting international events (biggest being the 13th World Festival of Youth and Students held in July 1989), accepting foreign students into their universities, interactions with pro-North Korean or Juche academic associations (World Conference on the Juche Idea), interacting with fan clubs such as the Korean Friendship Association, marketing of North Korean merchandise, and more recently, social media. Yet none of the above initiatives led to any major benefits for the DPRK's international status let alone gaining substantive support for its cause.

In the post-Cold War era, the DPRK became increasingly reliant on China not only to pursue more opportunities but also to fill the gaps left behind by USSR. Symbolically, the relations are about the "blood relations" from the Korean War, yet in substance, the partnership centers on post-Cold War economic and strategic interests. For the DPRK, China is not only the most essential economic partner but also a vital diplomatic guarantor. For China, the DPRK is the essential strategic buffer against an US ally and also a way to influence the state of affairs on the Korean peninsula. Indeed, there are questions over whether China would militarily defend the DPRK, and Pyongyang's military provocations and weapons tests have certainly frustrated Beijing. That said, Beijing has also buffered the harshest actions toward Pyongyang to maintain the status quo and has acted as a diplomatic guarantor. Likewise, the DPRK is often irritated by China's occasional cold attitudes and for taking part in the international sanctions against it, although Pyongyang recognizes that Beijing is not only the economic lifeline and dependable diplomatic umbrella but also a provider of modern technologies. Thus, despite the moments of unease between China and the DPRK, the bilateral ties have proven to be critical.

DPRK–Russia relations had chilled significantly in the 1990s, not simply due to the USSR's collapse but also because of Moscow's decision in 1995 to not extend the 1961 security treaty. Some changes were seen in February 2000 with Foreign Minister Igor Ivanov's visit Pyongyang that led to the signing of the Treaty of Friendship, Good-neighborliness, and Cooperation. In security, although the treaty merely stated that the two will not join any attacks against the other party, Pyongyang still values Moscow as a geopolitical counterweight against the US, and Russia wants to avoid instability and vulnerability of the DPRK for its own geopolitical interests. The DPRK–Russia relations are now much more about economic cooperation and logistics, notably in improving the transport links between the Russian Far East and the northeastern areas of North Korea, Russian access to North Korean materials and transport infrastructures, and the planned construction of pipelines via North Korea into South Korea. For the DPRK, Moscow forgave 90% of the debt Pyongyang accumulated from the Soviet years and provided access to some Russian technologies.

The DPRK's quest to strengthen the so-called international revolutionary forces has simply failed. The DPRK was conceived and raised during the Cold War, making the era, as Cha notes, the Stalinist state's "past, present, and future."[31] The post-Cold War world order has brought more misfortunes than benefits for Pyongyang, and it has struggled to find the optimum strategy to integrate and work with the fast-moving fluid environment while guaranteeing its survival. Even worse, the DPRK's bellicose reaction to the international community has led to a mass number of states taking part in international sanctions that led to further isolation and shunned economic opportunities. Thus, even though the DPRK currently has diplomatic ties with over 160 states, the amount of support for the DPRK cause, and Pyongyang's contribution to the international community has been more limited than ever.

National identity and strategic culture

Although the Three Revolutionary Forces for Reunification reveal *how* the DPRK views its security environment, it is national identity that shapes its perceptions and strategic culture. Much of the DPRK's national identity, however, is not formed by socialism, but rather by Kim Il-sung's values and visions based on ethno-centric nationalism packaged as Juche that was later mated with Kim Jong-il's Songun.

Pyongyang had never fully adhered to "pure socialism" from the outset. While the DPRK emulated the USSR's institutions and bureaucratic processes, they never embraced the Soviet style of socialism, to which Armstrong correctly described as the "'Koreanization' of Soviet communism, not the 'Sovietization' of North Korea."[32] Consequently, the North Korean communists were shaping the nation's identity based more on ethno-centric nationalism than Marxism–Leninism, much like many other Asian communist states that saw socialism as a vehicle for "national revival and modernity."[33] The loose foundations of socialism in Korea left niches for the WPK and the Kim dynasty to rewrite both the nationalistic and socialist narrative, leading to significant levels of revisionism. In the early years, the DPRK acknowledged the USSR's role in the liberation and the state's establishment, and portraits of Lenin and Stalin were seen in various places in North Korea. Yet as Kim Il-sung sculpted the monolithic system, references to the USSR were steadily displaced by Kim Il-sung's autonomous role in Korea's liberation and state construction. Likewise, the successive constitutional amendments steadily replaced references to Marxism–Leninism with Juche and Songun, finally ending up with Kimilsungism–Kimjongilism as the guiding principle of the party and the state.

Ethno-centric nationalism runs deep in the DPRK, even to the point where it justifies itself as the center of Korean civilization with the claimed tomb of Dangun just outside Pyongyang. Still, although the DPRK consistently stresses the importance of "Korean" values, there are caveats in understanding Pyongyang's conceptualization. Indeed, much is shared by the two Koreas when it comes to their "5000-year history" highlighted by countless conflicts, invasions, tributary status, occupations, and division that created the strong sense of strategic wariness and determination to establish a strong and independent sovereign state. However, Pyongyang and Seoul have fundamentally divergent views regarding the roots of the respective governments. On the one hand, the ROK claims that its government derives from the 1 March Independence Movement and the Provisional Government established in 1919. On the other hand, the DPRK claims its identity is based on Kim Il-sung's anti-Japanese partisan campaign and makes little references to the Provisional Government. Such meant that the DPRK's national identity and strategic perceptions are based on Kim Il-sung's *realpolitik* that formed through his personal experience and perceptions of the world.[34]

Given the nationalistic mood following liberation, Kim Il-sung's experience as an anti-Japanese partisan combined with his visions to create an independent state based on *realpolitik* gained much support, making him believe that he "owned the national narrative between leaders on the Korean peninsula."[35] Kim Il-sung went on to ideologically express the DPRK's national identity through Juche. Yet Juche proved to be an ideological amoeba without any clear meaning, leading to much debate among scholars. Scalapino and Lee described Juche as the "antithesis to Marxism" centered on nationalism.[36] Buzo described the nationalistic ingredient of Juche well, arguing that it is essentially about " 'purity' of all things Korean against the 'contamination' of all things foreign."[37] Cumings described Juche as an ideology that is centered on the leadership.[38] Similarly, Cha argues that Juche facilitated centralized control that practiced "collectivism in a Confucian rather than Marxist context."[39] All the above (and many others) are correct in that Juche was a tool to justify the centralized totalitarian Suryong leadership while shaping an independent national identity that is an antithesis to both liberal and socialist states.

Despite using the "Korean" label, however, Kim Il-sung's ideologies were far from authentic. First, Kim Il-sung utilized the features of various actors he had contact with in the past, including Stalinism, Confucian familism, nationalist ideas from Chondoism, Japan's emperor worship and total war mobilization as well as Christianity.[40] Second, there are clear double-standards in Kim Il-sung's claimed purity, given his time abroad and little association with local political movements in Korea prior to liberation.[41] But even though the use of various ideas and practices has made the DPRK's ideologies less authentic, they are consistent with Kim Il-sung's emphasis on selectively learning from other states by accepting the good and rejecting the bad.[42] Thus, the national identity engineered by Kim Il-sung was much less about advancing traditional and contemporary values, but more about creating an ideological cocktail that meets his agendas.

The DPRK took its national identity to new heights after the collapse of the USSR and the Eastern bloc. On the 47th anniversary of the WPK in October 1992, Kim Jong-il commented on the collapse of socialist states, criticizing the mismatches of Marxism–Leninism to current-day realities while justifying the DPRK's own socialist brand.[43] Thus, although the collapse of the USSR and the Eastern bloc was a major concern for the DPRK in the geopolitical and economic sense, it was also an opportunity to legitimize itself as the true and most resilient socialist power. The DPRK became increasingly self-justificatory, claiming to be the righteous and peaceful democratic revolutionaries who are fighting against the imperialistic, revisionist, war-mongering, reactionary infidels.[44] At the same time, the DPRK also began to define state power in its own ways, such as slogan of Kangsong Taeguk (strong prosperous state) that essentially boils down to legitimizing the regime under the concept of "our leader, our ideology, our military, and our political system."[45] While much is propaganda to legitimize the regime amid the challenging circumstances, such vaunts also reflect how Pyongyang saw itself as a lone and vulnerable power in the new post-Cold War geopolitical order.

The DPRK's national identity combined with the external and internal circumstances resulted in a peculiar form of combative, threat-based strategic culture with three notable aspects. The first is the mechanism designed to legitimize the leadership even under negative circumstances. The DPRK exploits the historical struggles (or the "arduous eras") of the Korean people as well as the various self-inflicted and self-imposed challenges since 1948 with one clear message – all the miseries are caused by the adversaries (internal and external) but the nation has prevailed to become an independent and mighty Juche state under the Kim dynasty's glorious and righteous leadership. Even to this day, Pyongyang argues that the state is constantly under the threat of being attacked and enslaved by the US and its allies and that the only way to survive and achieve the "revolutionary unification" is if the party, military, and the society sacrificially mobilizes and unites under the leadership. For the North Korean citizens, the Kim dynasty is the only option they know of. Thus, even if the domestic circumstances are unbearable and the far-reaching goals are almost impossible to achieve, the leadership has an infinite mandate to control and mobilize the masses.

The second is the use of the military as the catch-all means to solve all of the DPRK's problems and to complete the revolution. The fundamental ideas are linked to the past, where the DPRK argues that many of the Korean people's problems (i.e., Japanese colonization, division, threats from the US and its allies) were solved through armed struggles. For the DPRK, the KPA is not only the institution for defense but also the tool to achieve the leadership's raison d'état, as well as being the nation's identity and pride, and the key vehicle of the "revolution."[46] The military as the primary means of survival was much more pronounced under Songun, depicting the KPA as the role models of the society for their unwavering loyalty and dedication.[47] Minus the ideological rhetoric, the focus on the military has significantly shaped Pyongyang's strategic culture, where the law of the instrument – "when all you have is a hammer, everything looks like a nail" – applies. Consequently, the military proves to be the only functional leverage the state possesses, making every problem best solvable with the military.

The third is the strong sense of exceptionalism and isolation that justifies the DPRK's actions, namely those that ignore or contravene international agreements, laws, and norms. With the geopolitical events following the armistice, and more so in the post-Cold War era, the DPRK identified itself as a lone, vulnerable power that is constantly threatened by the US and its allies – further emboldening Pyongyang's already strong sense of Juche-ist *realpolitik*. Even if the DPRK is a member of the UN, it still stands by the belief that the international organ is exploited by the US. As a result, the DPRK has clung to the belief that it can only depend on itself and must do whatever it can to deal with the threats. Naturally, contradictions and double-standards emerge, where the DPRK criticizes the US for its possession of strategic arsenals and calls for denuclearization while justifying its own as an inevitable deterrent to defend itself. While dishonoring and manipulating deals indeed would affect its international reputation, Pyongyang believes that there is much less to lose than sticking to agreements and norms that would only constrain and make them more vulnerable.

Strategy: offensive or defensive?

The DPRK's ultimate goal remains to be the unification of the Korean peninsula under its terms. The question, however, is how the DPRK will try to achieve unification. The most extreme would be the complete communization of the Korean peninsula through force, but the realistic probability of successfully achieving this is narrow. Politically, Pyongyang will struggle to expand its political practices over into South Korea that has not only a larger population but also a greater diversity of political ideas and embracement of democracy. Economically, unification is unaffordable, given that Pyongyang struggles to manage even its own state economy. Militarily, while the DPRK may be able to inflict considerable damage and panic during the initial stages of conflict, it would be a matter of time before their strategies and regime will be burnt to ashes by the ROK and US forces. Such circumstances significantly lower the odds of successful and sustainable unification under DPRK terms.

As an interim alternative, the DPRK would push for the "one-state and two-systems" federation model. While seemingly more harmonious and peaceful, the model also presents opportunities for the DPRK to politically lure and molest the ROK into an arrangement under its control. Given the absolute support in North Korea for the regime, it would only take a few number of pro-North Korean elements in South Korea to form a majority that legitimizes the implementation of

Pyongyang's policies.[48] Moreover, the DPRK would also set conditions for the federation system, including an end to the ROK–US alliance, exit of US forces, and toning-down of ROK's defense posture to create exploitable opportunities for the DPRK to secure its control. Thus, the federation model can only be truly peaceful and successful under particular conditions, where the ROK does not compromise its identity, interests, and system based on liberal democracy.

The minimalist aim is the survival of the Kim dynasty. For the DPRK, the conditions for regime survival closely follows the Three Revolutionary Forces for Reunification. Above all, bolstering the Kim dynasty's authority in North Korea is of the utmost importance. Regarding the ROK, the DPRK regime's survival is better guaranteed with an administration that is not only is generous to the North but also distances itself from the US and Japan. Such a government will be a useful exit ramp but also a source for Pyongyang to milk out compromises and benefits. Then at the international level, the DPRK will count on likeminded states – particularly China and Russia – to use their leverage to buffer and deter external threats to the DPRK while also providing it with resources to keep them alive. Thus, the DPRK's aim of regime survival is not about maintaining the status quo, but rather to create conditions that minimize the disadvantageous circumstances while maximizing advantageous guarantees.

For the DPRK, it is not a matter of *either* regime survival or unification. After all, even the most strategically offensive states would not want to risk its own fate for expansionist ambitions. Rather, the question is about priority based on conditions at the time, where the DPRK will aggressively pursue unification under its terms when the circumstances are advantageous, while sticking to survival when the circumstances are disadvantageous. Some may opine that Pyongyang's rhetoric for unification is merely a domestic political communication tool, and that they are genuinely seeking to cooperate and make peace. Yet much of what the DPRK says about peace and reconciliation for unification is lip service to win the moral high-ground and lubricate dialogues for advantageous gains. In reality, the DPRK's terms for unification are arguably much more stringent than those of the ROK, where Pyongyang will only accept peace and unification that either takes place under their terms, or at the very least guarantees the preservation of the Kim dynasty. Unification in any other form is unacceptable, not only as it compromises the North Korean regime and its sovereignty but also the fate of political elites.[49] The layered strategic goals of unification and regime survival are therefore intertwined and symbiotic, where regime survival is the uncompromisable condition for unification, while unification under DPRK terms is the ultimate and surest way of survival.

Regardless of whether its unification or regime survival, there are common aspects. First, the DPRK's strategies exclusively focus on the Korean peninsula and the immediate periphery. Unlike its neighbors that are dependent on international trade and supply chains, the DPRK has limited offshore interests and has minimal concerns about the sea lines of communication (SLOC) beyond its own periphery. In security, other than the occasional deployment of small forces to the Middle East and Vietnam during the Cold War, the DPRK has refrained from becoming heavily involved in international security matters. The DPRK's expansionism is limited to its goal of unifying the Koreas under its terms, but nothing beyond. Even though the DPRK has indeed pursued strategic capabilities to strike the US and Japan, the purpose is to deter and defeat the threats to Pyongyang's interests on the Korean peninsula, as well as serving as a diplomatic card.

Second, whether it be for unification or regime survival, the DPRK seeks to change the regional geopolitical environment in ways that are favorable. On the high end, the DPRK sees a peace treaty and normalization of ties with the US as a substantive security guarantee, pushing to delegitimize the US–ROK alliance and the US presence on the Korean peninsula and ultimately the western Pacific. Achieving this would not only relieve the DPRK of the existential threat posed by the US and its allies but also open up opportunities to exploit the southern brethren and unify the Korean peninsula. At the same time, the DPRK is most sensitive to any developments in the defense strategies and readiness of the US, ROK, and Japan but also any signs of cooperation among the three. Historically, the DPRK has worked to decouple or at least create divisions in the US–ROK alliance but also between Seoul, Tokyo, and Washington to avoid confronting a trilateral pact. Moreover, the DPRK has been wary of any developments in relations the US, ROK, Japan has with China and Russia that would maximize their isolation and vulnerability. Thus, the bottom line is that the DPRK wants a divided region, as this is the environment that allows Pyongyang to maximize its leverage.

Third, whether it be for unification or regime survival, the DPRK has practiced offense as the best means of defense by depending heavily on the use of force. The CMCWPK has issued clear guidelines on the conditions for war, stating that it will declare war if: the ROK and US invade North Korea; nationalist (i.e., pro-DPRK) forces in South Korea requests intervention; internal and external circumstances become favorable for unification; and the ROK and US military provocation escalates and expands into full-scale conflict.[50] The use of force has been seen after the armistice, evidenced by the series of attempted armed attacks, assassinations, espionage, hijackings, infiltrations, seizure and sinking of ships, shooting down of aircraft, weapons tests, continued modernization of the armed forces, and others. Counterstability has also been a vital tool against South Korea to agitate the radical left and gain support while delegitimizing the US–ROK alliance and anyone against the DPRK. The offensive use of force is limited not only to achieve and secure unification but also to deter and ultimately neutralize the US, ROK, and Japan, revealing how the DPRK perceives the military as the only effective leverage to achieve its goals and to protect itself.

For the DPRK, the circumstantial developments through the lens of the Three Revolutionary Forces for Reunification combined with its sense of national identity and strategic culture have molded and emboldened the regime's combative, threat-based strategic perceptions. The DPRK views that it constantly faces both external and internal threats, and that it can only count on itself not only to ensure its security but also to achieve its strategic goals. Despite all the rhetoric, Juche was duly practiced, leading to the DPRK's quest to build and manage its own armed forces under the Line of Self-Reliant Defence doctrine as the means to deal with the threats against the regime.

Notes

1 According to Tsukamoto, the agreement also allowed the Soviet submarines to use North Korean ports. Katsuichi Tsukamoto, *kitachousengunto seiji [The North Korean Army and Politics]* (Tokyo, Japan: Hara Shobo, 2000), 34.
2 Ibid., 37–41.
3 See: Ibid., 51–52.
4 See: Il-sung Kim, "nongchongyeongrieui geumhubaljeoneul wihan uri dangeui jeongchaeke gwanhayeo [On Our Party's Policy for the Future Development of Agriculture] (3 November 1954)," in *Kim Il Sung jeojakjib [Kim Il Sung Works]*, vol. 9 (Pyongyang, DPRK: Joseon Rodongdang Chulpansa, 1980).
5 See: Il-sung Kim, "joguktongilwieobeul silhyeonhagi wihayeo hyeokmyeongryeokryangeul baekbangeuro ganghwahaja [Let Us Strengthen the Revolutionary Forces in Every Way so as to Achieve the Cause of Reunification of the Country] (27 February 1964)," in *Kim Il Sung jeojakjib [Kim Il Sung Works]*, vol. 18 (Pyongyang, DPRK: Joseon Rodongdang Chulpansa, 1982).
6 For detailed analysis of Kim Il-sung, see: Dae-sook Suh, *Kim Il Sung: The North Korean Leader* (New York, NY: Columbia University Press, 1988).
7 See: Adrian Buzo, *Politics and Leadership in North Korea: The Guerilla Dynasty*, 2nd ed. (London, UK and New York: Routledge, 2018), 13; Tsukamoto, *kitachousengunto seiji [The North Korean Army and Politics]*, 18–22.
8 Buzo, *Politics and Leadership in North Korea: The Guerilla Dynasty*, 31–37.
9 For a thorough analysis of the purges in the 1950s, see: Andrei N. Lankov, *Crisis in North Korea: The Failure of De-Stalinization, 1956* (Honolulu, HI: University of Hawaii Press, 2005).
10 Robert A. Scalapino and Chong-sik Lee, *Communism in Korea* (Berkeley, CA: University of California Press, 1972), 602–15.
11 Shunji Hiraiwa, *kitachousen: henbouwo tsuzukeru dokusaikokka [North Korea: The Continuously Transforming Dictatorship]* (Tokyo: Chuokoron Shinsha, 2013), 56.
12 Buzo, *Politics and Leadership in North Korea: The Guerilla Dynasty*, 73.
13 See: Kim, "dangwondeulsokeseo gyegeupgyoyangsaeobeul deouk ganghwahalde daehayeo [On Intensifying Class Education for Party Members] (1 April 1955)."; Il-sung Kim, "sasangsaeobeseo gyojojueuiwa hyeongsikjueuireul toechihago juchereul hwangribhalde daehayeo [On Eliminating Dogmatism and Formalism and Establishing Juche in Ideological Work] (28 December 1955)," in *Kim Il Sung jeojakjib [Kim Il Sung Works]*, vol. 9 (Pyongyang, DPRK: Joseon Rodongdang Chulpansa, 1980).
14 See: Il-sung Kim, "joseonminjujueuiinmingonghwaguk sahoijueuigeonseolgwa namjoseon hyeokmyeonge daehayeo [On Socialist Construction in the Democratic People's Republic of Korea and the South Korean Revolution] (14 April 1965)," in *Kim Il Sung jeojakjib [Kim Il Sung Works]*, vol. 19 (Pyongyang, DPRK: Joseon Rodongdang Chulpansa, 1982), 50–52.
15 See: Il-sung Kim, "gukgahwaldongeui modeun bunyaeseo jaju, jarib, jawieui hyeokmyeongjeongsineul deouk cheoljeohi guhyeonhaja [Let Us Embody the Revolutionary Spirit of Independence, Self-Sustenance and Self-Defence More Thoroughly in All Branches of State Activity] (16 December 1967)," in *Kim Il Sung jeojakjib [Kim Il Sung Works]*, vol. 21 (Pyongyang, DPRK: Joseon Rodongdang Chulpansa, 1983).

16 See: Jong-il Kim, "songunhyeokmyeongroseoneun uri sidaeeui widaehan hyeokmyeongroseonimyeo uri hyeokmyeongeui baekjeonbaekseungeui gichiida [The Songun-Based Revolutionary Line is a Great Revolutionary Line of Our Era and an Ever-Victorious Banner of Our Revolution] (29 January 2003)," in *Kim Jong Il seonjib [Kim Jong Il Selected Works]*, vol. 21, Expanded ed. (Pyongyang, DPRK: Joseon Rodongdang Chulpansa, 2013).
17 Dae-keun Yi, *bukhanguneun woae kudetareul haji ana [Why Don't the Korean People's Army Make a Coup]* (Paju, ROK: Hanul Academy, 2003), 98–104.
18 Charles K. Armstrong, *The North Korean Revolution, 1945–1950* (Ithaca, NY: Cornell University Press, 2003), 243–44.
19 See: Hiraiwa, *kitachousen: henbouwo tsuzukeru dokusaikokka [North Korea: The Continuously Transforming Dictatorship]*, 61.
20 Tsukamoto, *kitachousengunto seiji [The North Korean Army and Politics]*, 78.
21 See: Il-sung Kim, "joseoninmineui minjokjeokmyeongjeol 8.15haebang 15dolgyeongchukdaehoieseo han bogo [Report at the 15th Anniversary Celebration of the August 15 Liberation, a National Holiday of the Korean People] (14 August 1960)," in *Kim Il Sung jeojakjib [Kim Il Sung Works]*, vol. 14 (Pyongyang, DPRK: Joseon Rodongdang Chulpansa, 1981).
22 For detailed accounts of the inter-Korean interactions in the early 1970s, see: Don Oberdorfer, *The Two Koreas: A Contemporary History* (New York: Basic Books, 2001).
23 See: Daniel Wertz, JJ Oh, and In-sung Kim, "DPRK Diplomatic Relations," in *National Committee on North Korea Issue Brief* (Washington, DC: National Committee on North Korea, August 2016), 3.
24 See: Joseon Jungang Tongshinsa, *joseon jungang nyeongam 1963 [Korea Central Yearbook 1963]* (Pyongyang, DPRK: Joseon Jungang Tongshinsa, 1963), 159.
25 Tsukamoto, *kitachousengunto seiji [The North Korean Army and Politics]*, 134–36.
26 Ibid., 139.
27 Andrei N. Lankov, *The Real North Korea: Life and Politics in the Failed Stalinist Utopia* (Oxford, UK: Oxford University Press, 2013), 78.
28 See: Kim, "banjebanmitujaengeul ganghwahaja [Let Us Intensify the Anti-Imperialist Anti-US Struggle] (12 August 1967)." in *Kim Il Sung jeojakjib [Kim Il Sung Works]*, vol. 21 (Pyongyang, DPRK: Joseon Rodongdang Chulpansa, 1983).
29 Buzo, *Politics and Leadership in North Korea: The Guerilla Dynasty*, 83.
30 See: Wertz et al., "DPRK Diplomatic Relations," 6–7.
31 Victor D. Cha, *The Impossible State: North Korea, Past and Future*, 1st ed. (New York, NY: Ecco, 2012), 20.
32 Armstrong, *The North Korean Revolution, 1945–1950*, 241.
33 Lankov, *The Real North Korea: Life and Politics in the Failed Stalinist Utopia*, 7–8.
34 Buzo, *Politics and Leadership in North Korea: The Guerilla Dynasty*, 47–49.
35 Cha, *The Impossible State: North Korea, Past and Future*, 34.
36 Scalapino and Lee, *Communism in Korea*, 868.
37 Buzo, *Politics and Leadership in North Korea: The Guerilla Dynasty*, 56.
38 Bruce Cumings, "Corporatism in North Korea," *The Journal of Korean Studies* 4 (1982): 289.
39 Cha, *The Impossible State: North Korea, Past and Future*, 40.
40 For detailed discussions on the mixed aspects of Kim Il-sung and North Korea's political identity, see: Brian R. Myers, *The Cleanest Race: How North Koreans See Themselves and Why it Matters* (Brooklyn, NY: Melville House, 2010); Scalapino and Lee, *Communism in Korea*; Armstrong, *The North Korean Revolution, 1945–1950.*
41 Buzo, *Politics and Leadership in North Korea: The Guerilla Dynasty*, 42.

42 See: Kim, "joseonminjujueuiinmingonghwaguk sahoijueuigeonseolgwa namjoseon hyeokmyeonge daehayeo [On Socialist Construction in the Democratic People's Republic of Korea and the South Korean Revolution] (14 April 1965)," 51.
43 See: Jong-il Kim, "hyeokmyeongjeokdanggeonseoleui geunbonmunjee daehayeo [On the Fundamentals of Revolutionary Party Building] (October 10, 1992)," in *Kim Jong-il seonjib [Kim Jong Il Selected Works]*, vol.13 (Pyongyang, DPRK: Joseon Rodongdang Chulpansa, 1998).
44 See: Deok-sung Jeon, *songun jeongchie daehan rihae [Understanding Songun Politics]* (Pyongyang, DPRK: Pyongyang Chulpansa, 2004), 15.
45 See: Joint Editorial, "widaehan suryeongnim tansaeng 90dolseul matneun olhaereul kangsong taeguk geonseoleui saeroun biyakeui haero bitnaeija [Glorify this Year that Greets the 90th Birthday of President Kim Il Sung as a Year of a New Surge in the Building of a Powerful Nation]," *Rodong Sinmun*, January 1, 2002.
46 For the DPRK's description of the KPA during the formative years, see: Kwon Ri, *yeonggwangseureoun joseoninmingun [The Glorious Korean People's Army]* (Pyongyang, DPRK: Joseon Rodongdang Chulpansa, 1948).
47 See: Yi, *bukhanguneun woae kudetareul haji ana [Why Don't the Korean People's Army Make a Coup]*, 105–8. For examples of the DPRK's propagation of the KPA contributions to the society, see: Joseon Inmingun Chulpansa, *bulpaeeui hyeokmyeongmuryeok yeongungjeok joseoninmingun [The Invincible, Revolutionary and Glorius Korean People's Army]* (Pyongyang, DPRK: Joseon Inmingun Chulpansa, 2004).
48 Jang-yop Hwang, *birok gonggae: eodumeui pyeoni doin haetbyeoteun eodumeul balkghil su eobda [Sunshine Siding with Darkness Cannot Beat Darkness]* (Seoul, ROK: Wolgan Chosun, 2001).
49 As Lankov notes, North Korean elites feel there is "nothing to be gained and much to be lost in unification with the South." Lankov, *The Real North Korea: Life and Politics in the Failed Stalinist Utopia*, 118.
50 Yonhap News, "buk, jeonsiseonpogwon 'choigosaryeonggwan' – 4gae gigu gongdong myeongryeong [North Korea: Authority to Declare War Shifts from Supreme Commander to Joint Order by Four Institutions]," *Yonhap News* (22 August 2013), www.yna.co.kr/view/AKR20130822201200014.

References

Armstrong, Charles K. *The North Korean Revolution, 1945–1950*. Ithaca, NY: Cornell University Press, 2003.

Buzo, Adrian. *Politics and Leadership in North Korea: The Guerilla Dynasty*. 2nd ed. London, UK and New York, NY: Routledge, 2018.

Cha, Victor D. *The Impossible State: North Korea, Past and Future*. 1st ed. New York, NY: Ecco, 2012.

Cumings, Bruce. "Corporatism in North Korea." *The Journal of Korean Studies* 4 (1982): 269–94.

Hiraiwa, Shunji. *kitachousen: henbouwo tsuzukeru dokusaikokka [North Korea: The Continuously Transforming Dictatorship]*. Tokyo: Chuokoron Shinsha, 2013.

Hwang, Jang-yop. *birok gonggae: edodumeui pyeoni doen haetbyeoteun eodumeul bakghil su eobda [Sunshine Siding with Darkness Cannot Beat Darkness]*. Seoul, ROK: Wolgan Chosun, 2001.

Jeon, Deok-sung. *songun jeongchie daehan rihae [Understanding Songun Politics]*. Pyongyang, DPRK: Pyongyang Chulpansa, 2004.

Joint Editorial. "widaehan suryeongnim tansaeng 90dolseul matneun olhaereul kangsong taeguk geonseoleui saeroun biyakeui haero bitnaeija [Glorify this Year that

Greets the 90th Birthday of President Kim Il Sung as a Year of a New Surge in the Building of a Powerful Nation].” *Rodong Sinmun*, January 1, 2002.

Joseon Inmingun Chulpansa. *bulpaeeui hyeokmyeongmuryeok yeongungjeok joseoninmingun [The Invincible, Revolutionary and Glorius Korean People's Army]*. Pyongyang, DPRK: Joseon Inmingun Chulpansa, 2004.

Joseon Jungang Tongshinsa. *joseon jungang nyeongam 1963 [Korea Central Yearbook 1963]*. Pyongyang, DPRK: Joseon Jungang Tongshinsa, 1963.

Kim, Il-sung. “dangwondeulsokeseo gyegeupgyoyangsaeobeul deouk ganghwahalde daehayeo [On Intensifying Class Education for Party Members] (1 April 1955).” In *Kim Il Sung jeojakjib [Kim Il Sung Works]*. Vol. 9. Pyongyang, DPRK: Joseon Rodongdang Chulpansa, 1980a.

———. “nongchongyeongrieui geumhubaljeoneul wihan uri dangeui jeongchaeke gwanhayeo [On Our Party's Policy for the Future Development of Agriculture] (3 November 1954).” In *Kim Il Sung jeojakjib [Kim Il Sung Works]*. Vol. 9. Pyongyang, DPRK: Joseon Rodongdang Chulpansa, 1980b.

———. “sasangsaeobeseo gyojojueuiwa hyeongsikjueuireul toechihago juchereul hwangribhalde daehayeo [On Eliminating Dogmatism and Formalism and Establishing Juche in Ideological Work] (28 December 1955).” In *Kim Il Sung jeojakjib [Kim Il Sung Works]*. Vol. 9. Pyongyang, DPRK: Joseon Rodongdang Chulpansa, 1980c.

———. “joseoninmineui minjokjeokmyeongjeol 8.15haebang 15dolgyeongchukdaehoieseo han bogo [Report at the 15th Anniversary Celebration of the August 15 Liberation, a National Holiday of the Korean People] (14 August 1960).” In *Kim Il Sung jeojakjib [Kim Il Sung Works]*. Vol. 14. Pyongyang, DPRK: Joseon Rodongdang Chulpansa, 1981.

———. “joguktongilwieobeul silhyeonhagi wihayeo hyeokmyeongryeokryangeul baekbangeuro ganghwahaja [Let Us Strengthen the Revolutionary Forces in Every Way so as to Achieve the Cause of Reunification of the Country] (27 February 1964).” In *Kim Il Sung jeojakjib [Kim Il Sung Works]*. Vol. 18. Pyongyang, DPRK: Joseon Rodongdang Chulpansa, 1982a.

———. “joseonminjujueuiinmingonghwaguk sahoijueuigeonseolgwa namjoseon hyeokmyeonge daehayeo [On Socialist Construction in the Democratic People's Republic of Korea and the South Korean Revolution] (14 April 1965).” In *Kim Il Sung jeojakjib [Kim Il Sung Works]*. Vol. 19. Pyongyang, DPRK: Joseon Rodongdang Chulpansa, 1982b.

———. “banjebanmitujaengeul ganghwahaja [Let Us Intensify the Anti-Imperialist Anti-US Struggle] (12 August 1967).” In *Kim Il Sung jeojakjib [Kim Il Sung Works]*. Vol. 21. Pyongyang, DPRK: Joseon Rodongdang Chulpansa, 1983a.

———. “gukgahwaldongeui modeun bunyaeseo jaju, jarib, jawieui hyeokmyeongjeongsineul deouk cheoljeohi guhyeonhaja [Let Us Embody the Revolutionary Spirit of Independence, Self-Sustenance and Self-Defence More Thoroughly in All Branches of State Activity] (16 December 1967).” In *Kim Il Sung jeojakjib [Kim Il Sung Works]*. Vol. 21. Pyongyang, DPRK: Joseon Rodongdang Chulpansa, 1983b.

Kim, Jong-il. “songunhyeokmyeongroseoneun uri sidaeeui widaehan hyeokmyeongroseonimyeo uri hyeokmyeongeui baekjeonbaekseungeui gichiida [The Songun-Based Revolutionary Line is a Great Revolutionary Line of Our Era and an Ever-Victorious Banner of Our Revolution] (29 January 2003).” In *Kim Jong Il seonjib [Kim Jong Il Selected Works]*. Vol. 21. Expanded ed. Pyongyang, DPRK: Joseon Rodongdang Chulpansa, 2013.

———. "hyeokmyeongjeokdanggeonseoleui geunbonmunjee daehayeo [On the Fundamentals of Revolutionary Party Building] (October 10, 1992)," in *Kim Jong-il seonjib* [Kim Jong Il Selected Works], Vol.13. Pyongyang, DPRK: Joseon Rodongdang Chulpansa, 1998.

Lankov, Andrei N. *Crisis in North Korea: The Failure of De-Stalinization, 1956*. Honolulu, HI: University of Hawaii Press, 2005.

———. *The Real North Korea: Life and Politics in the Failed Stalinist Utopia*. Oxford, UK: Oxford University Press, 2013.

Myers, Brian R. *The Cleanest Race: How North Koreans See Themselves and Why it Matters*. Brooklyn, NY: Melville House, 2010.

Oberdorfer, Don. *The Two Koreas: A Contemporary History*. New York, NY: Basic Books, 2001.

Ri, Kwon. *yeonggwangseureoun joseoninmingun [The Glorious Korean People's Army]*. Pyongyang, DPRK: Joseon Rodongdang Chulpansa, 1948.

Scalapino, Robert A., and Chong-sik Lee. *Communism in Korea*. Berkeley, CA: University of California Press, 1972.

Suh, Dae-sook. *Kim Il Sung: The North Korean Leader*. New York, NY: Columbia University Press, 1988.

Tsukamoto, Katsuichi. *kitachousengunto seiji [The North Korean Army and Politics]*. Tokyo, Japan: Hara Shobo, 2000.

Wertz, Daniel, J. J. Oh, and In-sung Kim. "DPRK Diplomatic Relations." In *National Committee on North Korea Issue Brief*. Washington, DC: National Committee on North Korea, August 2016.

Yi, Dae-keun. *bukhanguneun woae kudetareul haji ana [Why Don't the Korean People's Army Make a Coup]*. Paju, ROK: Hanul Academy, 2003.

Yonhap News. "buk, jeonsiseonpogwon 'choigosaryeonggwan' – 4gae gigu gongdong myeongryeong [North Korea: Authority to Declare War Shifts from Supreme Commander to Joint Order by Four Institutions]." *Yonhap News*, August 22, 2013. www.yna.co.kr/view/AKR20130822201200014.

3 Command and control

While states vary in their institutional structures and processes, almost all exercise a centralized form of command and control over defense planning and also military operations. The rationale is clear – to ensure that the armed forces are formed and function strictly according to the government's policies and strategies. However, in authoritarian states where the governing party and/or the leader stand above the state institutions and processes, a multi-dimensional command and control mechanism is in place. The DPRK leadership has a three-dimensional, centralized, and politicized command and control over the defense planning process and also the armed forces. In defense planning, the CMCWPK and SAC play vital but contextualized roles, with the former planning and prescribing the policies, while the latter implementing and administering them. In operational command and control, the SCAF is in charge of the KPA as well as the paramilitary and reserve forces. The WPK leadership's command and control of the armed forces is reflected by not only the KPA Party Committee system but also the structure of the Ministry of National Defence (MND) (formerly the Ministry of National Security (MNS) between 1948 and 1972, and Ministry of People's Armed Forces (MPAF) between 1972 and 2020) where special organs such as the General Political Bureau (GPB), General Staff Department (GSD), Reconnaissance General Bureau (RGB), and the Military Security Command (MSC) play particularly important roles that make them more powerful than the ministry itself. Moreover, military-centric command and control is applied to internal security institutions, with the police institution known as the Ministry of Social Security (MSoS) (formerly the Ministry of Internal Affairs (MIA) between 1948 and 1962, and Ministry of People's Security (MPS) between 2000 and 2020) and the secret police known as the Ministry of State Security (MSS) led by senior KPA cadres.

The establishment of the armed forces

According to the DPRK, the KPA is directly connected to Kim Il-sung's anti-Japanese partisan movement. On 6 July 1930, Kim Il-sung united the various communists and anti-Japanese forces to form the Korean Revolutionary Army in Yitong County, China. Then on 25 April 1932, Kim Il-sung formed the Anti-Japanese People's Guerrilla Army that would become the Korean People's

Revolutionary Army (KPRA) two years later. The KPRA's legacies were carried on in the immediate post-liberation period, and after building the military education institutions, recruiting and training military cadres and personnel, the KPA was finally established on 8 February 1948. North Korean official accounts, however, grossly simplifies and distorts Kim Il-sung's military background and the complex story behind the KPA's establishment. While Kim Il-sung did lead a group of Korean partisans, they were working under, or at least in coordination with foreign forces. In 1931, Kim Il-sung joined the guerilla forces of the Chinese Communist Party known as the Northeastern People's Revolutionary Army that was later reorganized as the Northeast Anti-Japanese United Army. In 1940, Kim Il-sung and his comrades fled to the USSR, where he attended the Khabarovsk Infantry Officer School and served in the Soviet 88th Special Brigade until the Korean peninsula was liberated.

After liberation, complex developments were taking place in both sides of the 38th parallel, where various vigilantes formed to ensure local order. Yet on 12 October 1945, the USSR ordered the disarmament and disbandment of all armed groups, essentially restricting the development of security institutions to those led by Kim Il-sung. Significant developments followed, starting with the establishment of the Security Corps on 1 November 1945 that later became the provincial law enforcement organs.[1] In the following year, foundational developments for the military institution were underway with the establishment of the Railroad Security Corps on 11 January 1946 to guard the railway infrastructures that transported goods to and from the USSR.[2] On 15 August 1946, the Railroad Security Corps was expanded and reorganized to become the Security Cadre Training Battalion, later becoming the much larger and better armed People's Collective Forces on 17 May 1947.[3] Kim Il-sung also worked to establish educational institutions to train the officers who will lead the military institution. The first military school was the political Pyongyang Institute (later renamed the Second Officers School) in February 1946 followed by the more technical Central Security Officers Training School (later renamed the First Officers School) established in July 1946.[4]

The foundational developments for the air force began with the establishment of an aviator training school known as the Sinuiju Air Unit on 25 October 1945 that was reorganized as the Korean Aviation Association in the following month with branches in Chongjin, Pyongyang, Hamhung, and Hoeryong.[5] The Korean Aviation Association was later integrated into the People's Collective Forces with cadres trained at the Pyongyang Institute.[6] Recruitment for the air force was undertaken differently from the ground forces given the need for technical fluency, even recruiting some who were educated and trained in Japan in the pre-liberation years.[7] In training, approximately 900 experienced aviators were sent to USSR for re-training.[8] Moreover, given the number of aviators returning from Japan, there was much focus on political training to ensure conformity to the WPK. Later, the air unit was integrated into the KPA as its air arm to become the KPAAF.

The establishment of the naval forces was in less sync with the development of the KPA. On 5 June 1946, North Korea established the Maritime Security Force with fleet commands in Nampho and Wonsan. Then in August 1946, the central command was shifted from Wonsan to Pyongyang and the Maritime Security Force was renamed as the Coast Guards in December.[9] Although the cadres of the Maritime Security Force and Coast Guards were initially alumni of the Pyongyang Institute or Central Security Officers Training School, North Korea established the Maritime Security Officers Training School at Wonsan in June 1947.[10] Initially, the maritime forces operated as a law enforcement force under the directorate of the MIA, and it was not until August 1949 when it came under the command of the MNS to later become the KPAN.

On 4 February 1948, North Korea established the National Security Bureau under the People's Committee as a prelude to the inauguration of the KPA on 8 February 1948. Naturally, Kim Il-sung was at the top of the chain of command as the Chairman of the People's Committee of the DPRK, while his close comrade Choe Yong-gon was appointed as the Commander-in-Chief of the KPA. Once the state was established in September 1948, the National Security Bureau and the KPA command headquarters became the MNS directed by Choe Yong-gon.

Regarding recruitment, North Korea recruited local peasants and workers but excluded those who served in the Imperial Japanese Army and Navy and also the Manchukuo Imperial Army with the small exception of some specialists such as aviators and technicians who were educated and/or trained in Japan.[11] Still, there were major questions concerning the Korean communists who belonged to (or influenced by) movements other than Kim Il-sung's partisans. On the one hand, being exclusive would constrain the ability to build a large force. On the other hand, being inclusive could dilute Kim Il-sung's authority in the military. Kim Il-sung was helped by the Soviets who filtered the Korean returnees from China with military background. For example, Mao Zedong in October 1945 ordered the 4,000-strong Korean Volunteer Army led by Mu Jong to return to northern Korea via Sinuiju, but the USSR only allowed them to enter as disarmed individuals.[12] Furthermore, as Pyongyang prepped itself for the Korean War, about 30,000 Koreans who fought for the communists in the Chinese Civil War returned to North Korea.[13] Yet although the Korean communists of other origins were eventually accepted into the KPA, most were assigned as field commanders while the top echelons of the DPRK's defense planning and command and control were dominated by Kim Il-sung's close associates.

Given the lack of experience in establishing, developing, and operating a genuine military institution, as well as the time pressures Kim Il-sung faced, the Soviet tutelage was critical. During the formative years, the USSR had approximately 3,000 military advisors working in North Korea, and Pyongyang also sent around 10,000 personnel to Siberia to train with the Soviets.[14] Combined with Kim Il-sung's personal experiences in the Red Army, it was only natural for the KPA to be influenced by the Soviet military's organizational structure, processes, and culture.

Despite the contradictions between the DPRK's official accounts and actual developments, Kim Il-sung and his disciples – most notably An Kil, Choe Yong-gon, Kang Kon, and Kim Chaek – undoubtedly played leading roles in building the military institution and garnered most of the authoritative positions in the KPA. Yet Kim Il-sung's authority was also somewhat limited, as the DPRK was still arguably a loose authoritarian state, and the degree of party-centric politicization over the KPA was still moderate. Nevertheless, over the following decades, series of significant transformations took place, increasingly centralizing and politicizing the command and control of the KPA and the defense planning process under the Kim dynasty.

The defense planning organs

In the DPRK, there are three lines of command and control concerning military affairs. In operational command and control, the SCAF is at the helm. In defense planning, the CMCWPK sets the plans and prescriptions while the SAC is responsible for administration and implementation. Hence, although the leader is at the top, the three lines of command and control play particular roles. The North Korean party, state, and military governance system, however, is extremely complex, where various scholars have attempted to precisely map out the DPRK's command and control structures and processes but with great difficulty. While the biggest impediment is the high degree of opaqueness, the other is the numerous changes to the structures and processes over the decades. Moreover, personnel changes have often taken place without official announcements, creating difficulties in capturing accurate snapshots of the personnel lineup of the key party, state, and military organs.[15]

The party

Although the Political Bureau of the CCWPK is the highest policy-making organ of the WPK, the CMCWPK is the party's planning and decision-making apparatus regarding military affairs. According to Article 29 of the WPK charter, the CMCWPK is the organ that "discusses and decides the WPK's military policies, directs projects to strengthen the KPA's military capability and develop the military industry, and commands the military." In essence, the CMCWPK can be best described as the core of the DPRK's defense planning, convening numerous meetings throughout the year to discuss and decide on comprehensive and detailed matters relating to strategies, operational and tactical concepts and doctrines, political affairs in the armed forces, R&D and procurements, and other matters that require major correction or reconfiguration.

Generally, the cast of the CMCWPK is elected at the CCWPK plenums, although the exact lineup has been unclear on many occasions. Almost always, the CMCWPK comprises the Director of the GPB, Minister of MND/MPAF, Chief of the GSD, Director of the GSD Operations Bureau, Director of the RGB, as well as the Military Affairs Department (MAD) and the Munitions Industry

Department (MID) of the CCWPK. Although less regular, on many occasions, the ministers of the MSS and MSoS, director of the General Rear Services Bureau, commanders of the KPAAF, KPAN, and KPASRF, and a number of other senior field commanders and military elders have also been included. The field commanders serving in the CMCWPK are based on the particular strategic priorities at the time. Moreover, CMCWPK meetings often take place in expanded form, including senior cadres from the armed forces, party, military industry, as well as the internal security organs to announce the strategies, policies, and orders.

The precise history of the CMCWPK remains unknown, although it is generally agreed that the senior party apparatus specialized for defense planning was conceived with the establishment of the MCCCWPK at the Fifth Plenary Meeting of the Fourth CCWPK in December 1962. The inaugural members of the MCCCWPK in 1962 are unknown, but the members in 1968 included: Kim Il-sung as Chairman; and Choe Hyon, Choe Yong-gon, Kim Chang-bong (Deputy Prime Minister; MNS, Minister), Kim Kwang-hyop (Deputy Prime Minister), Ri Yong-ho, and Sok San (MSoS, Minister) as members.[16] The establishment of the committee was about reforming the party under Kim Il-sung's absolute authority as well as establishing a better means of military management under the WPK leadership. The first agenda and achievement of the MCCCWPK was the issuance of the Line of Self-Reliant Defence and the establishment of the military-industrial complex that will later be coordinated by the Second Economic Committee (SEC). Thus, the establishment of the MCCCWPK was essentially about setting the institutional and bureaucratic foundations for the DPRK's Juche-ist defense planning.

Major changes took place when Kim Jong-il became the heir apparent at the Sixth WPK Congress in October 1980. The MCCCWPK expanded in size and included a larger cohort of both military and party cadres, now consisting of: Kim Il-sung as Chairman; and Choe Hyon, Choe Sang-uk, Jo Myong-rok, Jon Mun-sop, Ju To-il, Kim Chol-man, Kim Il-chol, Kim Jong-il, Kim Kang-hwan, O Jin-u (MPAF, Minister; GPB, Director), O Kuk-ryol (GSD, Chief), O Paek-ryong, O Ryong-bang, Paek Hak-rim, Ri Pong-won, Ri Tu-ik, Ri Ul-sol, and Thae Pyong-ryol as members.[17] Moreover, at the Sixth Plenary Meeting of the Sixth CCWPK in August 1982, the MCCCWPK was reorganized to become the CMCWPK, becoming an organ parallel with the CCWPK. Yet since, announcements on the CMCWPK membership and its activities were rarely announced until the 2010s. Much of the silence could be explained by the fact that the WPK did not hold any congresses or conferences during this period, and even the plenums that announce the membership of the CMCWPK had not been held since December 1993.

After Kim Il-sung's death, the CMCWPK is known to have comprised the following: Kim Jong-il (SCAF; NDC, Chairman) as *de facto* Chairman; and Choe Kwang (GSD, Chief), Choe Sang-uk, Jo Myong-rok (KPAAF, Commander), Jon Mun-sop, Ju To-il, Kim Chol-man, Kim Il-chol (KPAN, Commander), Kim Kang-hwan (GSD, Deputy Chief), O Jin-u (MPAF, Minister; GPB, Director), O Kuk-ryol, Paek Hak-rim, Ri Pong-won (GPB, Deputy Director), Ri Tu-ik,

Ri Ul-sol, and Thae Pyong-ryol as members.[18] Some reshuffles took place by the time the DPRK resumed its activities in 1997, particularly with the deaths of Choe Kwang, Ju To-il, O Jin-u, and Thae Pyong-ryol. The CMCWPK in 1997 now included: Kim Jong-il as *de facto* Chairman; and Jo Myong-rok (GPB, Director), Kim Il-chol (MPAF, Minister), Kim Myong-guk, O Ryong-bang, Paek Hak-rim, Pak Ki-so, Ri Ha-il, Ri Pong-won, Ri Ul-sol (Guard Command, Commander), and Ri Yong-chol as members.[19] Much is unknown, however, about the actual membership of the CMCWPK during the Kim Jong-il era. While it is hard to be precise with the absence of official announcements, some of the changes are calculable, given that some of the key positions that are consistently represented in the commission (e.g., ministers of MND/MPAF, directors of GPB, and chiefs of GSD).

The CMCWPK was publicly reinvigorated at the Plenary Meeting of the CCWPK in September 2010, announcing the lineup that included the following: Kim Jong-il as Chairman; Kim Jong-un and Ri Yong-ho (GSD, Chief) as Vice-Chairmen; and Choe Kyong-song, Choe Pu-il, Choe Ryong-hae, Choe Sang-ryo (Missile Guidance Bureau, Commander), Jang Song-thaek, Jong Myong-do (KPAN, Commander), Ju Kyu-chang (MID), Kim Jong-gak, Kim Kyong-ok, Kim Myong-guk (GSD Operations Bureau, Director), Kim Yong-chol (RGB, Director), Kim Yong-chun (MPAF, Minister), Kim Won-hong (GPB, Deputy Director), Ri Pyong-chol (KPAAF, Commander), U Tong-chuk (MSS, First Deputy Director), and Yun Jong-rin (Guard Command, Commander) as members.[20] The public revival of the CMCWPK was not only about refreshing and enhancing the DPRK's defense planning system under the WPK but also to use the organ as a platform to strengthen Kim Jong-un's authority. Naturally, Kim Jong-un was promoted to Chairman of the CMCWPK at the WPK Conference in April 2012. There were also other notable changes, particularly with the promotion of Choe Ryong-hae to Deputy Chairman, indicating his fast rise to become one of the most senior aides in the third-generation leadership. In addition, Hyon Chol-hae (MPAF, First Deputy Minister), Kim Rak-gyom (KPASRF, Commander), and Ri Myong-su (MPS, Minister) were also appointed as members.[21] More importantly, numerous changes took place in the forthcoming years due to the increasingly frequent changes in the top military positions.

The CMCWPK announced at the Seventh WPK Congress in May 2016 was more centralized, going without the position of Vice-Chairman. The members consisted of: Kim Jong-un as Chairman; Choe Pu-il (MPS, Minister), Hwang Pyong-so (GPB, Director), Kim Kyong-ok (GSD, First Deputy Director), Kim Won-hong (MSS, Minister), Kim Yong-chol (RGB, Director), Pak Pong-ju (Premier), Pak Yong-sik (MPAF, Minister), Ri Man-gon (MID, Director), Ri Myong-su (GSD, Chief), Ri Yong-gil (GSD, former Director), and So Hong-chang (MPAF, Deputy Minister).[22] Naturally, there have been some reshuffles due to changes in the commanders, directors, and ministers, particularly in the GPB, GSD, and the MND/MPAF.

The current CMCWPK was elected at the Eighth WPK Congress in January 2021, featuring a remarkable generational change in the lineup, including:

Kim Jong-un as Chairman; Ri Pyong-chol as Vice-Chairman; Jong Kyong-thaek (MSS, Minister), Jo Yong-won, Kang Sun-nam, Kim Jo-guk, Kim Jong-gwan (MND, Minister), Kwon Yong-jin (GPB, Director), O Il-jong (MAD, Director), O Su-yong (SEC, Chairman), Pak Jong-chon (GSD, Chief), Ri Yong-gil (MSoS, Minister), and Rim Kwang-il (RGB, Director) as members.[23]

The MAD and the MID of the CCWPK also deserve particular mention. The MAD functions as a thinktank for practical matters relating to the armed forces while also overseeing military-related affairs within the WPK.[24] Likewise, the MID handles matters relating to the military-industrial complex and manages the SEC. Although affiliated to the CCWPK rather than the CMCWPK, the two departments play vital roles in drafting, shaping, and appropriating the party's military policies that are then approved and ordered by the CMCWPK.

Another important party organ is the Organization and Guidance Department (OGD) that is arguably the most powerful organ within the CCWPK. The OGD is in charge of deciding and managing appointments, promotions and demotions, as well as dismissals of not only WPK members but also state and military officials. On top of personnel management, the OGD is the architect and executors of the WPK's policies while also directing the internal security organs such as the MSS to surveil and arrest any persons that are suspected of deviating from the regime. Although the OGD does not run the DPRK's defense planning, it nevertheless plays a pivotal role in shaping the military sector by overseeing the MND and the armed forces.[25]

The state

The SAC is the highest state organ that administers and implements orders relating to state affairs and defense planning. The SAC has particular powers with the MND, MSoS, and the MSS being directly subordinate to the commission as opposed to the cabinet. However, the SAC itself is young, and the state organ for defense affairs has undergone several changes since the 1950s.

The origins of the state institution specializing in defense affairs traces back to the Military Commission established under the Extraordinary Decree issued at the Presidium of the SPA on 26 June 1950. At the time of its foundation, the commission consisted of: Kim Il-sung as the Chairman; and Choe Yong-gon (MNS, Minister), Hong Myong-hui (Deputy Prime Minister), Jong Jun-thaek (State Planning Commission, Chairman), Kim Chaek (Deputy Prime Minister; Ministry of Industries, Minister), Pak Hon-yong (Deputy Prime Minister; Ministry of Foreign Affairs, Minister), and Pak Il-u (MIA, Minister) as members.[26] Little is known about the Military Commission after the armistice. Given the composition and timing of its establishment, it is possible that the Military Commission was designed to mobilize resources for the war effort, making it less relevant after the armistice. The other possibility is that the commission was being phased out by the MCCCWPK, particularly given the strengthening of the WPK's rule under Kim Il-sung during the 1950s and 1960s.

The state defense planning organ was reestablished when the DPRK adopted the revised constitution in December 1972, founding the NDC subordinate to the new and powerful CPC. At the time, the NDC focused purely on military-related matters and worked alongside other specialized commissions under the CPC, such as the Domestic Policy Commission, Foreign Policy Commission, and the Justice and Security Commission. When the NDC was launched, the membership was relatively compact, consisting of only four members: Kim Il-sung, Choe Hyon (MPAF, Minister), O Jin-u (GSD, Chief), and O Paek-ryong (CMCWPK, Vice-Chairman).[27] Yet similar to the CMCWPK, very little was publicized about the NDC for the next two decades, largely because all the orders were made by Kim Il-sung as the President as opposed to the chair of the NDC.

The 1990s saw major changes in the NDC's authority and structure. The NDC elected at the First Session of the Ninth SPA in May 1990 included the following: Kim Il-sung as Chairman; Kim Jong-il as First Vice-Chairman; Choe Kwang (GSD, Chief) and O Jin-u (MPAF, Minister; GPB, Director) as Vice-Chairmen; and Jon Pyong-ho, Ju To-il (Pyongyang Defence Command, Commander), Kim Bong-ryul, Kim Chol-man (SEC), Kim Kwang-jin, Ri Ha-il, and Ri Ul-sol as members.[28] Yet in April 1992, the constitution was amended to make the NDC an independent state organ. The changes were not simply about empowering the NDC, but in large part to promote Kim Jong-il's military leadership while keeping Kim Il-sung's party and state leadership. Under the newly amended constitution, the NDC Chairman or the SCAF no longer needed to be the President. Changes were made accordingly on 9 April 1993, with Kim Jong-il promoted to the position of NDC Chairman.

After the three years of mourning Kim Il-sung, the DPRK amended its constitution in 1998 that set the Kim Jong-il political system. The biggest change was the abolishment of the CPC and the position of President. Consequently, the NDC not only became the state institution that would have exclusive and direct power over the MPAF and other security institutions such as MSoS and MSS, but also extended its authority to areas beyond national defense, consequently becoming a mega organ handling broader state affairs. Furthermore, the promotion of the NDC was particularly important for Kim Jong-il and his military-centric leadership, to what Buzo correctly described as the "institutional expression of Songun."[29] Although the NDC appeared to be a less formal and permanent organ of state power, it proved to be extremely useful for Kim Jong-il to legitimize and strengthen his leadership over both the military and the state.

The NDC of the Tenth SPA elected in September 1998 revealed generational changes that set the tone for the Kim Jong-il era, consisting of: Kim Jong-il as Chairman; Jo Myong-rok (GPB, Director) as First Vice-Chairman; Kim Il-chol (MPAF, Minister) and Ri Yong-mu as Vice-Chairmen; and Jon Pyong-ho, Kim Chol-man, Kim Yong-chun (GSD, Chief), Paek Hak-rim (MSoS, Minister), Ri Ul-sol, and Yon Hyong-muk as members.[30] In part, the changes were due to the death of senior cadres such as O Jin-u in February 1995 and Choe Kwang in February 1997. The NDC of the Eleventh SPA elected in September 2003 saw some reshuffles with Ri Yong-mu and Yon Hyong-muk (replaced by Kim Yong-chun in April 2007) now serving as Vice-Chairmen, while Choe Ryong-su, Jon Pyong-ho, Kim Il-chol, and Paek Se-bong were elected as members.[31] One notable aspect was that the NDC now included three figures from the military industry sector – Jon Pyong-ho, Kim Chol-man, and Paek Se-bong – reflecting the prioritization of the strategic weapons program.

At the First Session of the Twelfth SPA in April 2009, the constitution was amended again, further empowering the NDC as the "supreme defense leadership body of State power." Further changes were seen in the NDC of the Twelfth SPA which now included: Kim Jong-il as Chairman; Jo Myong-rok (GPB, Director) as First Vice-Chairman; Kim Yong-chun (MPAF, Minister), O Kuk-ryol, and Ri Yong-mu as Vice-Chairmen; and Jang Song-thaek, Jon Pyong-ho, Ju Kyu-chang, Ju Sang-song (MPS, Minister), Kim Il-chol, Kim Jong-gak (GPB, First Deputy Director), Paek Se-bong, and U Tong-chuk as members.[32] Yet the NDC of the Twelfth SPA was frequently disrupted by personnel changes with the dismissal of Kim Il-chol in May 2010 and death of Jo Myong-rok on 6 November 2010, leading to the promotion of Jang Song-thaek to Vice-Chairman on 7 June 2010 and appointment of Pak To-chun as member on 7 April 2011.[33] More significant, the death of Kim Jong-il in December 2011 left the position of Chairman of the NDC vacant until the appointment of Kim Jong-un as the "First Chairman" at the Fifth Session of the Twelfth SPA in April 2012. In addition, there were new faces in the membership, with the appointment of Choe Ryong-hae (CMCWPK, Vice-Chairman; GPB, Director), Kim Won-hong (MSS, Minister), and Ri Myong-su (MPAF, Minister).[34] Further changes were seen in 2013 due to ministerial changes, with Kim Jong-gak replaced by Kim Kyok-sik as Minister of MPAF, and Ri Myong-su replaced by Choe Pu-il as Minister of MPS.[35]

The first "full" NDC of the Kim Jong-un era was announced at the First Session of the Thirteenth SPA in April 2014. The new NDC saw some changes with Choe Ryong-hae promoted to Vice-Chairman, while Jang Jong-nam replaced Kim Kyok-sik as the Minister of MPAF and Jo Chun-ryong as a representative from the military industry sector. The NDC of the Thirteenth SPA was affected by the frequent personnel changes. Just weeks after the election of the NDC, Hwang Pyong-so replaced Choe Ryong-hae as the director of the GPB.[36] In June, the Minister of MPAF Jang Jong-nam was replaced by Hyon Yong-chul who was later reportedly executed in 2015.[37] Then in September 2014, former KPAAF commander Ri Pyong-chol who assumed an unknown but key role in the CCWPK was elected as a member.[38] However, the NDC of the Thirteenth SPA was short-lived, as the constitutional amendment adopted at the Fourth Session of the Thirteenth SPA on 29 June 2016 abolished the NDC and created a new organ known as the SAC.

The establishment of the SAC was based on two factors. The first was about completing the transition from Kim Jong-il to Kim Jong-un by reconfiguring the state management system under the third-generation leadership. The second was about ironing out the ambiguous nature of the NDC, particularly as Songun no longer needed to be as emphasized like in the 1990s and 2000s. Both rationales were much about the fact that the NDC, as correctly described by Madden, had "outlived its organizational life cycle."[39] Thus, the establishment of the SAC was coming, particularly with Kim Jong-un's emphasis on formalizing and personalizing his leadership while also modernizing the DPRK. More changes came in April 2019 with the constitutional amendment that set the Chairman of the SAC as the official head of state – a position that was not formalized during the Kim Jong-il era.

The most notable aspect of the SAC is the composition that includes a more diverse cross-section of the state government. While the NDC was largely made up of military, law enforcement, intelligence, and military industry elites, the SAC now included the Prime Minister and Foreign Minister, as well as other senior party and state officials. Yet despite some of the notable changes in the structure, it is premature to claim that the SAC is demilitarized and desecuritized in nature than the NDC. Above all, the SAC inherited some of the key features of the NDC, most notably with the direct administration and prioritization of the MND/MPAF, MSoS/MPS, and MSS. Moreover, within the SAC is the National Defence Committee, indicating the commission's full administrative control over military-related affairs. Even regarding the membership, the reduction in the number of military cadres and increase in the number of other state and party officials were much more about exercising a more "all-in" approach to carry out Pyongyang's strategic objectives rather than diluting the military's role. Given that the strategic weapons had developed significantly, greater attention was now on how those assets will be utilized by the state in foreign policy and inter-Korean relations. Thus, the replacement of the NDC with the SAC was much about creating a more robust national security council-type of organ under Kim Jong-un's leadership as opposed to some kind of demilitarizing reform.

The first SAC elected at the SPA included the following: Kim Jong-un as Chairman; Choe Ryong-hae, Hwang Pyong-so (GPB, Director), and Pak Pong-ju (Premier) as Vice-Chairmen; and Choe Pu-il (MPS, Minister), Kim Ki-nam, Kim Won-hong (MSS, Minister), Kim Yong-chol, Pak Yong-sik (MPAF, Minister), Ri Man-gon (OGD, First Deputy Director), Ri Su-yong (Ministry of Foreign Affairs, former Minister), and Ri Yong-ho (Ministry of Foreign Affairs, Minister) as members.[40] Some changes were made in 2018, with figures including Jong Kyong-thaek (MSS, Minister), Kim Jong-gak (GPB, Director), and Thae Jong-su (MID) elected as members.[41] At the First Session of the Fourteenth SPA in April 2019, the SAC was joined by Choe Son-hui (Ministry of Foreign Affairs, First Vice Minister), Kim Jae-ryong, Kim Su-gil (GPB, Director), and No Kwang-chol (MPAF, Minister). Further changes took place at the Third Session of the Fourteenth SPA in April 2020, with Kim Hyong-jun, Kim Jong-gwan (MPAF, Minister), Kim Jong-ho, Ri Pyong-chol, and Ri Son-gwon (Ministry of Foreign Affairs, Minister) elected as members.[42] The changes in the SAC membership since 2016 were not simply due to positional changes but also about renewed and reemphasized strategies to deal with military affairs and the US. Further changes are expected in future SPA sessions, particularly in reflecting the changes introduced at the Eighth WPK Congress in January 2021 and other personnel reshuffles, as well as the potential for Kim Jong-un to adopt a more powerful position as the head of state to command the SAC.

Supreme Commander of the Armed Forces

The SCAF is in charge of the operational command and control of both the KPA and the KPISF, as well as the other paramilitary and reserve forces. The SCAF is not an institution in itself, but a position for the leadership to issue operational commands and instructions to the armed forces as well as making other announcements such as the promotion of generals. The authority of the SCAF depends much on the level of war readiness ordered. According to Shin and Kim, the DPRK's war readiness alert is divided into five levels: (1) war situation, (2) semi-war situation, (3) combat mobilization, (4) standby for combat mobilization, and (5) combat awareness.[43] Thus, the higher the status of alert, the SCAF has greater authority in both operational command and control of not only the KPA but also the state and the party.[44]

When the KPA was first established, the operational command and control was tasked to Choe Yong-gon who was the then Minister of MNS. However, the position was reorganized soon after the outbreak of the Korean War. On 4 July 1950, the Supreme Command of the KPA was established under the decree of the Presidium of the SPA, making Kim Il-sung the commander-in-chief of the armed forces. Kim Il-sung's exclusive right to serve as the SCAF was confirmed with the adoption of the new constitution in December 1972, stating that the SCAF is to be served by the President. Then on 24 December 1991, Kim Jong-il was named the SCAF, which led to changes in the constitution in 1992 to allow the Chairman of the NDC to serve the command position by default. Then in 2016, the constitutional clauses relating to the SCAF were changed again, where the position is served by the Chairman of the SAC. Thus, the SCAF has been tailored according to the leadership transitions to ensure that only the Kim dynasty can serve the position.

While the SCAF is served by the leader (or the heir apparent during the final stages of leadership transition) by default, the appointment process after Kim Il-sung has been carried out by the WPK. Kim Jong-il was appointed on 24 December 1991 at the Nineteenth Plenary Session of the Sixth CCWPK, and Kim Jong-un was appointed on 30 December 2011 by the Political Bureau of the CCWPK. While the appointments of the successors were well expected, the procedures highlight once again how the WPK is at the center of key decisions relating to the armed forces.

Organs within the MND and KPA

The decisions and orders from the CMCWPK and SAC are implemented by the MND. The MND contains various bureaus and departments in charge of: education and training, finances, intelligence and counterintelligence, logistics, operations, personnel, planning, and other administrative affairs.[45] However, the GPB,

GSD, RGB, and the MSC are connected to, and work under particular authorities given their particular importance to the leadership – political affairs, operational command, external security intelligence, and counterintelligence. The importance of the four organs is evidenced by their representation in the CMCWPK, where the directors and chiefs of the GPB and GSD have often assumed more senior positions in the commission than the ministers of the MND. Therefore, it is more accurate to describe the MND as an umbrella institution that functions as administrative and logistical caretakers of the armed forces.

For several decades, the MND/MPAF along with the MSoS and the MSS has worked under a special hierarchy instead of being affiliated to the cabinet along with the other ministries. In April 1982, the MPAF was separated from the Cabinet to become directly subordinate to the CPC in 1986. Then after the abolishment of the CPC in September 1998, the MPAF came under the direct control of the NDC and later the SAC in June 2016. Although this makes the MND one of the most privileged bureaucratic organs in the DPRK, it is also exposed to the greatest amount of control by the leadership. The purpose of the special hierarchy is not only to ensure that the security apparatuses work in strict accordance with the leadership but also to facilitate smoother and more expedited means of implementing plans and orders.

While the MND is a state organ, it is strongly connected to the WPK given the power of the CMCWPK and the organs of the CCWPK including the MAD, MID, and the OGD. To ensure that tasks are carried out in strict accordance with the WPK, there are OGD and Propaganda and Agitation Department officials who are attached to the MND.[46] Moreover, the senior cadres of the MND are members of the WPK, making it natural for the military to function as a property of the party.

The WPK leadership's tight control over the armed forces is evidenced by the KPA Party Committee system, with party committees established in the corps, divisions, and regiments, while cells and subcommittees are established in the companies and platoons. The WPK committees, cells, and subcommittees are branched into two lines, one being the GSD and the other being the GPB. The party committee system within the KPA was established after the armistice and the intra-party struggles during the mid-1950s. To ensure that the KPA is under full control of the WPK leadership, Kim Il-sung called for the establishment of a party committee system that brings together the GPB and GSD cadres.[47] While the system could be described as a mechanism to harmoniously fusion practical and political elements under the WPK's lines, much emphasis is on political discipline and loyalty, thus essentially making it a check-balance mechanism to ensure the military's unwavering allegiance to the WPK leadership.

General Political Bureau

The WPK's power over the KPA is enabled by the GPB that standardizes the identity and ideological texture of the armed forces. While the GPB is technically subordinate to the MND, they work under the strict command and control of the OGD to ensure that the military conforms to the WPK.

Before the KPA was established, political education in the formative military organizations took place in the form of "cultural training" conducted by the Pyongyang Institute. When the KPA was founded in February 1948, the Cultural Training Bureau was established within the armed forces and was headed by Kim Il-sung's close comrade Kim Il. Yet while there certainly was significant level of bias toward the WPK, the Cultural Training Bureau was less absolutist regarding Kim Il-sung, propagating broader socialist and nationalistic values. The broad approach reflected the politics at the time, as Kim Il-sung had not yet secured his dictatorship of the party and state, forcing him to take a more balanced approach. Yet much changed with the outbreak of the Korean War, particularly when the KPA was being pushed back by the UN forces. On 21 October 1950, Kim Il-sung called for the establishment of WPK organs within the military to deal with the decline in morale and discipline, leading to the establishment of the GPB.[48]

Significant developments in the GPB came in the post-armistice years, particularly as Kim Il-sung sought to ensure that the military does not fall into the hands of political rivals or become influenced by ideological alternatives. As Kim Il-sung worked to purge the Soviet and Yanan factions within the military in the late 1950s, he called on to correct and strengthen the GPB's role in the KPA to ensure integrity and loyalty to the WPK.[49] More developments came in the 1960s after Kim Il-sung purged military hardliners, prompting Kim Il-sung to further expand the GPB's authority by appointing commissars and officers to the lower levels of the KPA.[50] Ostensibly, the developments were about Kim Il-sung's calls to strengthen ideological education and discipline in the KPA for the so-called Juche revolution. Yet in reality, the measures were designed to ensure the military's absolute loyalty to the leader. The successors to Kim Il-sung also focused much on the role of the GPB, particularly during times of leadership transition and major socio-economic crises. For instance, in the 2000s, Kim Jong-il argued that the GPB officers were becoming irresponsible, calling for robust measures to ideologically strengthen the KPA.[51] While the leader's argument was that the military must be disciplined and united for the so-called revolution, it was entirely about curbing any deterioration in discipline and loyalty that would undermine the regime's control over the armed forces.

The WPK's top–down political control of the KPA is enabled by the very structure of the GPB, with political commissars represented in the committees and departments above the battalion level while political directive officers work in the cells and subgroups at the company and platoon levels. The role of these political commissars and officers can be described as political administrators, educators, enforcers, chaplains, and caretakers all rolled into one.[52]

The GPB has a number of departments and sections that play critical roles. In organizational management, the Organization Department is tasked with personnel management to assess the profiles of personnel and make decisions on appointments, transfers, and promotions that are then administered by the MND Cadre Bureau.[53] The Organization Department is also in charge of managing the party membership of KPA personnel. While the KPA is a property of the WPK, party membership is limited to the senior officers and/or those with good political credentials. The non-WPK members make up most of the KPA but are members of the Kimilsungist–Kimjongilist Youth League (formerly the Kim Il Sung Socialist Youth League) run by the CCWPK. Personnel who perform well in military and political activities are then assessed by the Organization Department on their suitability to join the WPK. Furthermore, the GPB monitors the behavior of personnel, where any infringements and questionable actions are reported to the OGD by the Report Section, and later dealt with by the MSC.[54]

The other critical department is the Propaganda Department in charge of ideological indoctrination through political activities, including various campaigns, conferences, and rallies. The GPB leads campaigns such as the Gold Star Elite Guard at the division and brigade levels, the O Jung-hup Seventh Regiment at the regiment levels, and the Three Revolution Red Flag at the company levels.[55] Moreover, much of the contents of major internal publications of the KPA are determined and vetted by the GPB, issuing literatures aimed at uplifting loyalty and morale as well as educating personnel on code of conduct and discipline.

Given the significant level of authority held by the GPB, it is not surprising that the political officers and commissars enjoy a prestigious status in the KPA and the North Korean society. A tenure in the GPB provides significant advantages for individuals aspiring to join the highest echelons in the KPA and the WPK. Naturally, joining the GPB is far from easy, with stringent selection criteria based on political credentials and loyalty than practical military aptitudes.[56] Once commissioned, GPB personnel work separately from the rest of the KPA undertaking significantly less military training than the field personnel and are also generally on a faster promotion track.[57] However, there have been some special cases in the past. For example, in the 1970s when the KPA was growing in size, a number of senior WPK bureaucrats were transferred to the GPB to fill shortages in the political commissars and officers.[58]

The director of the GPB has always held a senior post in the CMCWPK and the SAC (and NDC), often serving as deputy chairman. Particularly, as the significance of the GPB grew from the 1960s, the GPB director essentially became the representative face of the KPA aside from the SCAF. In most cases, the directors of the GPB have had long careers as political officers and commissars, with many of them having deeper careers as WPK bureaucrats as opposed to the KPA. Recent examples of GPB directors with almost minimal levels of military experience include Choe Ryong-hae and Hwang Pyong-so. On some occasions, however, some field officers became directors of GPB, such as the case of Ho Bong-hak, Jo Myong-rok, Kim Jong-gak, and O Jin-u. The appointment of the field officers to the GPB top seat was not simply based on military credentials but rather because they were one of the most trusted KPA cadres for the DPRK leadership, particularly in the case of Jo Myong-rok and O Jin-u who played vital roles when Kim Il-sung and Kim Jong-il worked to bolster their control over the KPA.

General Staff Department

The GSD is the operational arm of the MND tasked to manage the various KPA units as well as the reserve and paramilitary forces in wartime. Although the GSD is technically affiliated to the MND, it has its own direct means of communication to the SCAF.

At the heart of the GSD is the Operations Bureau that is linked to the various commands and divisions of the KPA. The Operations Bureau is broken into 11 offices that are each in charge of specific areas, including general administration, commands from the SCAF, development of operational and tactical plans, education and training, supervision of commands, and affairs relating to the Military Armistice Commission.[59] KPA units under the Operational Bureau include the 21 corps and divisions of the ground forces, the KPAN, KPAAF, KPASOF, and KPASRF. In addition, paramilitary organs such as the KPISF and the reserve forces come under the Operational Bureau in wartime. Generally, the head of the Operations Bureau serves as the First Deputy Chief of GSD.

In addition to the Operations Bureau, the GSD is known to have a number of directorates to support the KPA's operations in areas including, but not limited to: communications and signals, education and training, engineering, construction, transport and logistics, topography, and so forth. Little information, however, is available on the actual numbers and the specific names of the bureaus and offices affiliated to the GSD, sometimes leading to confusion and discrepancies. For instance, Lee Yeong-hoon who formerly served as a GSD officer claims that there are approximately 65 directorates that are part of the MND and GSD.[60] Thus, it is unclear as to whether the aforementioned bureaus and offices other than the Operations Bureau are subordinate to the GSD, or other bureaus of the MND (such as the General Rear Services Bureau, General Planning Bureau, Construction Bureau, Transport Bureau, and Education Bureau).

The Chief of the GSD has its own position in the CMCWPK. However, there seems to be no specific regulations concerning the appointment of the Chief of the GSD as the position has been served by cadres of varying ranks (mostly Vice Marshal or General). Like many other senior positions in the military sector, the Chief of the GSD has served irregular terms, where some such as O Jin-u and Kim Yong-chun served for more than ten years while some only served less than 12 months. Moreover, it has not been rare for the Chief of the GSD to be more senior than the MND. On several occasions, some who previously served ministerial positions were later appointed as the Chief of the GSD. As recent examples, Ri Myong-su and Kim Kyok-sik who previously served as Ministers of the MSoS and MND, respectively, were later appointed as the Chief of the GSD.

Although the GSD is not responsible for political activities, it is nonetheless involved in political affairs as the department is tied to the WPK committees and cells of the KPA to work with the GPB. Moreover, the GSD is exposed to the GPB's political activities and oversight to ensure the operational arm and the units work in absolute consistency with the WPK.

Reconnaissance General Bureau

The RGB is the intelligence arm of the MND in charge of collecting information and conducting other forms of clandestine operations against ROK, Japan, and the US, as well as illicit trade activities. The RGB was conceived in 2009 as a merger of the WPK's political intelligence organs and the GSD Reconnaissance Bureau. The merger cleaned up some of the overlaps and stove-piping that was endemic to the existence of multiple intelligence organs.[61] While there are still other intelligence organs operated by the DPRK, the RGB nonetheless is considered to be *the* external intelligence organ serving the leadership.[62]

Detailed information about the RGB is little available, although some analysts have noted that the bureau consists of six departments in charge of various affairs, including espionage and other covert operations, intelligence analysis, logistics, and training.[63] The activities of the RGB are overseen by the director who is often represented in the CMCWPK. The first director of the RGB was Kim Yong-chol who now serves as one of the right-hand figures of Kim Jong-un. As Kim Yong-chol moved into more senior roles in the leadership, the head of the RGB has been succeeded by Jang Kil-song in September 2017 who was later replaced by Pak Kwang-il in December 2019, and then by Rim Kwang-il in May 2020.

Military Security Command

The MSC is the counterintelligence and policing agency within the KPA, tasked with investigating, screening, and arresting military personnel suspected of committing conspiracies, corruption, and other infringements, as well as ensuring information security. Moreover, the MSC is believed to deal with threats to the leadership not only within the military but also in the general populace, indicating coordination or even overlap with the state's internal security organs such as the MSoS and the MSS.

Much is unknown about the recruitment and training of MSC personnel, although it is presumed that the procedures are extremely stringent given the nature of the duties. Those recruited go on to receive education and training at specialized universities. Once commissioned, MSC officers are then deployed to work undercover in the various units of the KPA where they would watch the activities of questionable personnel and their families, eavesdrop into communications, check on relations between the GSD and GPB personnel, as well as investigating any irregular activities or events.[64]

The decision-making framework

Over the decades, Kim Il-sung, Kim Jong-il, and Kim Jong-un took every opportunity to strengthen the leadership's multi-dimensional command and control mechanism over both defense planning and the armed forces. Although the DPRK does not have a completely parallel party-state defense planning organ like the Central Military Commission in China, it must be borne in mind that the SAC (and previously the NDC) has never been on competitive terms with the WPK. The CMCWPK and SAC play different roles in defense planning, where the former focuses on planning and prescribing the military's plans and strategies, while the latter administers and implements the former's orders. The purpose is clear, to ensure the leadership has full command and control over the DPRK's defense planning and the armed forces. Moreover, the key senior members of the CMCWPK and the SAC serve in the Political Bureau of the CCWPK, which again demonstrates the close connection between the party and management of military affairs.

Indeed, during the Kim Jong-il era, there were times when the DPRK placed more public emphasis on the NDC and the SCAF. However, one explanation is that Kim Jong-il declared a "semi-war" alert status in March 1993 which seemed to have prolonged for some time, thereby leading to the frequent issuance of orders as the SCAF. Moreover, much had to do with Kim Jong-il's reconfiguration of the state institutions, where he abolished the CPC and instead utilized the military management organ (i.e., NDC) as more effective means to exercise executive authority. The reason for the military-centered state governance was two-fold. One was that the final and most critical stages of the transition to Kim Jong-il was taking place amid challenging external and internal circumstances, leading to the idea that rule through military-centric martial law would be the most plausible option. The other was to keep the military firmly under Kim Jong-il's control to curb any risks of the military working on their own accord, or worse, coup d'état.

Still, there is little evidence to suggest that the NDC (let alone the KPA) outsized the authority of the WPK. Although there was a greater number of military cadres in the Political Bureau of the CCWPK under Kim Jong-il that made them the most influential strategists and policymakers, the military never had direct access to the institutional levers over the WPK. Even during the Kim Jong-il years, the CMCWPK was active, issuing key directives such as Order Number 002 in April 2004 that set guidelines for war mobilization.[65] The WPK's consistent power over the NDC and the KPA (and other state institutions for that matter) is also evident in the management of personnel. At the time of the NDC, the lineup of the commission was determined by the CCWPK and CMCWPK, which was then "elected" by the rubber-stamp SPA.[66] One can assume the same procedures apply to the SAC.

The most striking aspect about the DPRK is how the centralized system was sustained amid the generational transitions. After setting the new Kim Il-sung-based constitution in 1972, the DPRK has amended the Socialist Constitution six times (until the time of writing) with each of them designed to legitimize the dynasty-based succession. The constitutional amendments were necessary as the three leaders used different platforms to establish their leadership, making the two successions take place in distinctive fashion. The constitutional amendments in 1992 and 1998 empowered and legitimized Kim Jong-il's leadership as the NDC Chairman and as the SCAF without being the President. As for Kim Jong-un, the amendments in 2012, 2016, and 2019 had justified his succession and his state leadership that is now based on the SAC. Thus, the constitutional amendments were made to legitimize the successor while also ironing out any contradictions created by the transition from one generation of the leadership to the next. Likewise, one of the key purposes of the amendments to the WPK memorandum was to further legitimize and strengthen the leadership's authority.

Successions in the DPRK have never been a simple linear process of one leader to the next, but one of hierarchical parallel inheritance. The constitutional amendment in 1998 abolished the position of President and made Kim Il-sung the "Eternal President of the Republic." Similarly, Kim Jong-il became the "Eternal General Secretary of the WPK" and also the "Eternal Chairman of the NDC" after his death, and Kim Jong-un officially succeeded the leadership as the "First Secretary of the WPK" and "First Chairman of the NDC." Thus, even if the predecessors are physically dead, they are politically kept alive with emeritus titles. The purpose is to ensure a consistent continuum in the dynasty. Although the incumbent holds absolute power, their legitimacy is based on the values, visions, and prophecies of the predecessor. For example, Kim Jong-il argued that the best way to show loyalty to Kim Il-sung was to follow the Songun ideology. By the same token, Kim Jong-un has also referred to the legacies and visions of Kim Il-sung and Kim Jong-il, calling for further modernization to continue the "revolution." Hence, the inheritance of the leadership was not simply about relaying the policies and visions of the predecessor but also essentially about embodying and enlarging the system inherited by the successor.

Centralization amid succession also depended heavily on the senior cadres that have occupied the key party and state positions relating to defense planning. In many states, senior bureaucratic positions are rotated on a regular basis following certain procedures based on merit, rank, and stages of their careers. In the DPRK's case, the nature of appointments changed over time. At least until the late 1950s, the management of senior posts relating to military affairs was certainly bias to Kim Il-sung's partisan faction but was much more routine and bureaucratic in nature. Yet as Kim Il-sung bolstered his totalitarian authority, the senior echelons of Pyongyang's defense planning were exclusively limited to the most trusted cadres handpicked by the leader. The figures in the highest echelons are not ideological fanatics that blindly follow the leader, but those who are loyal and able to give well calculated and planned proposals despite the harsh strategic realities.

The leadership's tight control over the military is enabled by the massive cohort of general/admiral-grade officers that provides a large pool of senior cadres who are assignable to key posts. According to the ROK Ministry of Unification, the KPA currently has six Vice Marshals, 28 Generals, 44 Colonel Generals, 107 Lieutenant Generals, and 1,128 Major Generals.[67] The numbers cannot simply be explained by the size of the KPA. One explanation for the overwhelming cohort of generals and admirals is the fact that officers of senior ranks do not have a mandatory retirement age. Moreover, while in many countries the top positions in the military are often the final assignments for one's career, this is not the case in the DPRK, with many who served as the head of the GPB, GSD, MND/MPAF, MSoC, MSS, or other command positions staying on to play advisory roles, or even returning to key portfolios. Thus, it is common to have senior cadres who are well over their 70s, or even into their 80s. But the other explanation is the number of WPK bureaucrats who are politically commissioned to senior military ranks. For example, all senior cadres of the OGD (Section Chief and above) hold ranks of Lieutenant General and above.[68] Similarly, many other senior cadres of the WPK and sometimes the state are also given senior military ranks. Kim Jong-il's funeral in December 2011 was a good example, where some top-level party cadres such as Jang Song-thaek, Ju Kyu-chang, Kim Kyung-hui, and Pak To-chun were seen donning military uniforms. Although one could argue that the actual roles of such figures in the KPA are politically symbolic and do not have any command roles, they nonetheless play vital roles in the decision-making process – particularly those who are appointed to the CMCWPK and NDC/SAC. Moreover, the promotion of cadres to general-grade ranks is particularly important in the leadership transition context where the successors authorized mass promotions upon assuming their leadership not simply for human resource reasons but to nurture a cohort of KPA leaders with absolute loyalty to the leadership.

Among the many senior military cadres, a number of them have played particularly important roles. The figures can be categorized according to generations based not simply by their age, but when they joined the military system. The first generation consists of Kim Il-sung's comrades from the anti-Japanese partisan campaign and those who closely worked with the leader during the formative years, such as An Kil, Choe Hyon, Choe Yong-gon, Ho Bong-hak, Jon Mun-sop, Kang Kon, Kim Chaek, Kim Ik-hyon, Kim Il, Kim Ryong-yon, Nam Il, O Jin-u, O Paek-ryong, Paek Hak-rim, Pak Song-chol, Ri Jong-san, Ri Ul-sol, Rim Chun-chu, and others. Many of these figures worked with Kim Il-sung during the formative years to form the nucleus of the KPA leadership and also for some the DPRK regime itself. Some of the members of the Manchurian partisan faction either died or were purged between the 1950s and 1960s. But those who survived played instrumental roles in managing and strengthening the KPA in the 1960s and onward.

One noteworthy figure was O Jin-u, Kim Il-sung's loyal protégé and arguably the most senior military figure in the post-armistice period (other than members of the Kim dynasty). O Jin-u shot up the military ranks from the KPA's formative years and his rise accelerated after the purges in the mid- and late-1950s.[69] O Jin-u was appointed as the Director of GPB in 1967 (until his death in 1995), Chief of GSD in 1968 (until 1976), and then the Minister of MPAF in 1976 (until his death in 1995). To this day, O Jin-u remains to be the only known cadre who served as the head of the GPB, GSD, and the MPAF. O Jin-u was one of Kim Il-sung's most trusted aides, playing a vital role from the 1960s in not only shaping and managing the KPA strictly according to the leader's political and strategic outlook but also spearheading the purging of some of the military's top brass that were alleged to have contradicted the leadership.

Another important figure from the first generation is Choe Hyon who closely worked with Kim Il-sung from the anti-Japanese partisan years. Choe Hyon was one of the most experienced guerillas in the Manchurian faction, having led some key battles against the Japanese forces, but his lack of education had constrained him from holding any significant political posts during the early years of the regime and instead took on field command positions in the KPA. Still, Choe Hyon's closeness and loyalty to Kim Il-sung had helped him to rise up the ranks in both the party and the military during the 1960s, being appointed to the Minister of MNS in 1968 to which he served until 1976. Like O Jin-u, Choe Hyon not only played a pivotal role in shaping the KPA but also was one of the figures that proactively endorsed Kim Jong-il's succession.

The second generation of elders is those who joined the KPA in the late 1940s and 1950s, including Hyon Chol-hae, Jo Myong-rok, Jang Song-u, Jon Jae-son, Kim Il-chol, Kim Tu-nam, Kim Yong-chun, O Kuk-ryol, Pak Ki-so, Ri Ha-il, Ri Pong-won, Ri Yong-mu, and others. Many of them are alumni of the elite Mangyongdae School for the Bereaved Children of Revolutionaries (now Red Flag Mangyongdae Revolutionary School), an elite education institution for the children of not only military and party elites but also some orphans. The second-generation cadres joined the senior echelons of the regime around the 1970s when they reached general-grade ranks.

The third generation, who now serve as the core of the Kim Jong-un regime are those who started their military or party careers from the 1960s or 1970s, including Choe Pu-il, Choe Ryong-hae, Hwang Pyong-so, Kim Jong-gwan, Kim Kyok-sik, Kim Su-gil, Kim Won-hong, Kim Yong-chol, O Il-jong, Pak Jong-chon, Ri Myong-su, Ri Pyong-chol, Ri Yong-gil, Ri Yong-ho, Won Ung-hui, and others. Many of these figures have served in the CMCWPK, SAC/NDC, MND/MPAF as well as the MAD and MID. Moreover, a number of the cadres from this category are the children of first-generation revolutionaries, such as the case of Choe Ryong-hae who is the son of Choe Hyon, and O Il-jong who is the son of O Jin-u. The third-generation cadres are special in that they were less exposed to the intra-party rifts in the 1950s and 1960s and were nurtured when the KPA and the WPK were under the full control of the leadership.

The elders have played central roles during the time of leadership transitions. When Kim Jong-il officially became the successor at the Sixth WPK Congress in October 1980, he was flanked by many of Kim Il-sung's trusted colleagues from the first- and second-generations. In similar fashion, Kim Jong-il also teed-up some of the most senior figures from both the second- and third-generation for Kim Jong-un. The figures served not only as key advisors and guardians in the CMCWPK and the NDC/SAC but also as enforcers to ensure that the KPA remains under control. Moreover, it is important to note that not all the trusted elders necessarily have official portfolios and not all those with portfolios are the most senior. Some of the elders with deep expertise and experience have worked behind the scenes to directly advise the leader. One classic example is O Kuk-ryol, who was occasionally appointed to positions in the NDC, but seems to have played a larger role as an unofficial advisor to Kim Jong-il and also Kim Jong-un.

Another notable feature regarding the DPRK's military management is the high degree of centralization but with limited levels of integration. The DPRK has worked to create a check-balance system among the components of the MND to deny any opportunities for organized challenges against the leadership. One obvious example of the check-balance system is the aforementioned KPA Party Committee system to ensure that the military conforms to party lines. Check-balance systems also exist in the decision-making echelons, where military cadres are categorized according to field officers, political, internal security (i.e., MSoS and MSS), and military industry officials. The check-balance system is also seen in the information security regimes that monitor and restrict communication among the MND components. Even for information, with the exception of general notices and publications, the type of information circulated within the military is strictly determined according to particular departments and positions.[70] Indeed, the restriction of information according to the levels of security clearance is common practice even in democratic states to ensure information security. Nevertheless, the level of the measures implemented in the DPRK is far more thorough and restrictive as they are more for the purpose of executing a check-balance system to prevent and surgically remove causes and symptoms of political challenges before they infect the whole armed forces and the regime.

The combination of the hyper-centralized structure and the high level of standardization was also pivotal in allowing the leadership to formulate policies and plans in a single, unopposed direction. Specifically, standardization was about smoothing the planning and decision-making processes according to the leadership's policies and visions while muting any lobbying. During the 1960s, KPA officers were purged not only for alleged disloyalty but also for voicing and pursuing interests different from the leadership. In January 1969, "hardliners" such as Minister of MNS Kim Chang-bong, GPB Director Ho Bong-hak, and Chief of GSD Choe Kwang were banished not only for the failed guerrilla attack on the Blue House in 1968 but also for their pursuit of high-tech military platforms that Kim Il-sung did not favor.[71] Instead, Kim Il-sung asserted that military modernization conforms to hybrid warfare that best matches with "Korean conditions." The problem was not so much about the technologies that were being proposed by the hardliners, but rather the fact that Kim Il-sung saw this as a potential for lobbying that would undermine his authority. This created a strong sense of political bias and groupthink that significantly impacted decisions concerning military capabilities and operations.

Care is also needed when looking at the centralized and politicized system. Although the leaders have always made the final decisions, they nonetheless delegate the military, party, and state cadres to draft the details of the plans. For example, the MND would make the requests for acquisitions or adjustments in the KPA that are checked by the OGD, processed by the MAD and MID, and then approved, and ordered by the CMCWPK.[72] The reverse is also true, where the leader would make specific demands or requests that are then processed by the MAD and MID, and then passed to the MND via the SAC for implementation. One could even make the argument that Kim Jong-il and Kim Jong-un have been particularly dependent on the senior cadres given their lack of military experience compared to Kim Il-sung. Still, the degree of independent, bottom–up influences are limited, as the cadres can only work within the confines of the doctrines and visions laid down by the leadership. Moreover, there is little room for error, as any failures could lead to demotion, dismissal, or in the worst case, execution of the cadres responsible.

Of course, there are limits to the actual amount of authority held by the senior cadres as they too are vulnerable to dismissals, demotions, and purges. Mass purges of senior KPA officers took place on three occasions, with the first taking place just before and after the armistice to blame the failed military campaign; in 1958 to rid those from the Soviet and Yanan factions; and in the late 1960s to remove the military "hardliners." Arguably, the biggest purge was the one in 1958 when around 85% of general-grade officers were reportedly banished.[73] In March 1958, Jang Pyong-san from the Yanan faction was banished for allegedly planning a coup d'état, while Kim Il-gyu and Kim Il-jong were purged for teaching in military colleges that the KPA belonged to the Democratic Front for the Unification of the Fatherland rather than the WPK.[74] Fears of factionalism and coup d'états were also emboldened by events outside of North Korea, including the 1961 and 1979 coups in the ROK, as well as the execution of Nicolae Ceausescu in Romania in 1989. The various events combined with the Kim leadership's own issues and fears led to major reconfiguration of the counterintelligence and internal security institutions in the DPRK such as the establishment of the MSS in 1973 and also further enhancing the MSC's role in the KPA. Although there were cases of attempted coups such as the "Frunze Incident" in 1992 and the "6th Army Corps Incident" in 1995, both were swiftly preempted before they materialized.

While dismissals, demotions, retirements, and purges often take place, there have been a number of cases where some cadres were later reinstated. For example, Choe Kwang was dismissed as Chief of GSD in March 1969 but was reinstated to both party and state positions several years later and was appointed to the Chief of GSD and Minister of MPAF. Likewise, O Kuk-ryol was dismissed as Chief of GSD in February 1988 but bounced back in the following year to serve as the director of the WPK Civil Defense Department and then the director of the WPK Operations Department. Later in 2009, O Kuk-ryol was appointed as the Vice-Chairman of the NDC and reportedly played a leading role in the establishment of the RGB.

One notable aspect about the Kim Jong-un leadership is the higher turnover of military cadres, with eight changes in the MND (out of 15 since 1948), seven for the Chief of GSD (out of 19 since 1948), and five for the Director of GPB (out of 17 since 1948) at the time of writing. Some of the most senior cadres, many of whom served as Kim Jong-un's regents during the transition process were demoted, purged, or retired by the very leader they were advising and guarding. While some of the purges and dismissals were relatively discreet, some have been extremely brutal, most notably with the execution of Jang Song-thaek in December 2013 and Hyon Yong-chol in April 2015. The simple explanation for the changes would be that Kim Jong-un was sensitive to the risks of becoming over-influenced by the regents or wedged in an internal power struggle. However, while some were genuinely banished, many were simply retired due to their age. The fast-aging and diminishing cohort of senior cadres that aided Kim Jong-un were vital in his succession but became an impediment in managing the military (and state affairs) to produce substantive results. Thus, it became imperative for

Kim Jong-un to take on a more rotational approach to establish the right combination of cadres to execute defense planning in a more effective manner while also maintaining maximum power distance with the elites.[75]

Impact on defense planning

While Kim Il-sung aimed to emulate Stalin during the formative years, he far outstripped his idol in the extent of centralization and politicization that resulted in the socialist dynasty. Over time, Kim Il-sung and his successors exploited and manipulated every opportunity to solidify their absolute rule. The leadership succession in the DPRK was never about replacement but about affirming, inheriting, preserving, and fine-tuning the centralized and politicized system. Indeed, not all was smooth, particularly with the intra-party rifts in the 1950s, and also during times of transition when there were growing levels of dissatisfaction against Kim Jong-il, and uncertainties concerning the third generation Kim Jong-un leadership. Still, the Kim dynasty prevailed, and the three generations continuously fine-tuned their centralized and politicized authority over the defense planning process and the armed forces. In fact, Kim Jong-un has arguably been the most effective, establishing the SAC and contextualizing the role of the WPK that has led to a more balanced and optimized command and control mechanism.

The centralized and politicized nature of the command and control system significantly impacted the decision-making process. The centralized and politicized system is beneficial in that it allows the leadership to expedite decisions and execution of policies. At the same time, the North Korean political system represents a prime example of path-dependence, creating major rigidities beyond the point of no return. As the famous French philosopher Émile-Auguste Chartier once said, "Nothing is more dangerous than an idea when it's the only one you have." The problems are not so much about ideologies, as they are merely flexible tools to bless the leadership's visions and actions. Rather, the problem is how the leadership's decisions are incarcerated by their end-state for regime security rather than development. The leadership's obsession with their own survival led to a system that has potent levels of nepotism and rigidity in decision-making, consequently breeding a cohort of cadres whose minds are dominated by their experience and allegiance to the leadership. Consequently, the decision-making process became less flexible, where any bold and innovative measures outside of the political lines would contradict and unwind the very system sculpted by the three generations of leaders. The path-dependent nature of the DPRK regime has created major issues for defense planning with the excessive but inevitable focus on regime survival coming at the expense of decisions for best practice. Although various prescriptions could be put forward to effectively improve the KPA's readiness, the measures must not in any way affect the regime's centralized and politicized command and control of the armed forces. Even if Kim Jong-un is taking on a more hands-on approach focusing on substantive results, the measures fall far short of reform let alone changes in strategies.

Notes

1 Katsuichi Tsukamoto, *kitachousengunto seiji [The North Korean Army and Politics]* (Tokyo, Japan: Hara Shobo, 2000), 15.
2 Ibid.
3 Ibid., 15–16, 18, 23.
4 See: Il-sung Kim, "pyongyanghakwongaewonsikeul chukhahayeo [Congratulations on the Opening of the Pyongyang Institute] (23 February 1946)," in *Kim Il Sung jeojakjib [Kim Il Sung Works]*, vol. 2 (Pyongyang, DPRK: Joseon Rodongdang Chulpansa, 1980).
5 Wan-gyu Choi, "joseoninminguneui hyeongseonggwa baljeon [The Formation and Development of the KPA]," in *bukhaneui gunsa [North Korean Military Affairs]*, ed. Bukhan Yeongu Hakhoi (Seoul, ROK: Gyeongin Munhwasa, 2006).
6 Tsukamoto, *kitachousengunto seiji [The North Korean Army and Politics]*, 23.
7 Ibid., 30.
8 Ibid.
9 Choi, "joseoninminguneui hyeongseonggwa baljeon [The Formation and Development of the KPA]," 21–22.
10 Ibid.
11 For instance, Ri Hwal who played a key role in establishing the Korean Aviation Association and later served as the deputy commander of the air battalion was a graduate of the Nagoya Aviation School.
12 See: Robert A. Scalapino and Chong-sik Lee, *Communism in Korea* (Berkeley, CA: University of California Press, 1972), 333–34; Tsukamoto, *kitachousengunto seiji [The North Korean Army and Politics]*, 14.
13 Min-ryong Lee, *Kim Jong-il chejeeui bukhangundae haebu [Anatomy of the Kim Jong-il Regime's North Korean Army]* (Seoul, ROK: Hwanggeumal, 2004), 325.
14 See: Tsukamoto, *kitachousengunto seiji [The North Korean Army and Politics]*, 28.
15 Effort has been made to identify the positions of the various personnel, but not all were able to be identified, and many remain to be unclear. The positions of key military cadres are provided in parentheses.
16 Scalapino and Lee, *Communism in Korea*, 935, 1381.
17 Teruo Komaki, "roudoutoudairokkaitaikaino toshi: 1980nenno chousenminsyusyugijinminkyouwakoku [The Year of the Sixth Congress of the Workers' Party of Korea: The Democratic People's Republic of Korea in 1980]," in *Ajia Doukou Nenpou 1981 [Annual Report on Trends in Asia 1981]* (Chiba, Japan: Institute of Developing Economies, 1981), 86.
18 Masahiko Nakagawa, "'idaina syuryou' no shikyo: 1994nenno chousenminsyusyugijinminkyouwakoku ["The Death of the Great Suryong": The Democratic People's Republic of Korea in 1994]," in *Ajia Doukou Nenpou 1995 [Annual Report on Trends in Asia 1995]* (Chiba, Japan: Institute of Developing Economies, 1995), 88; ROK Ministry of Unification, "bukhanjeongbopotal [North Korea Information Portal]," nkinfo.unikorea.go.kr.
19 Teruo Komaki, "'kunanno kougun' kade kimujonirutaiseigaseisikihossoku: 1997nenno chousenminsyusyugijinminkyouwakoku [The Official Inauguration of the Kim Jong-il Regime under the 'Arduous March': The Democratic People's Republic of Korea in 1997]," in *Ajia Doukou Nenpou 1998 [Annual Report on Trends in Asia 1998]* (Chiba, Japan: Institute of Developing Economies, 1998), 91; ROK Ministry of Unification, "bukhanjeongbopotal [North Korea Information Portal]".
20 Masahiko Nakagawa, "koukeitaiseikouchikuno junbi hajimaru: 2010nenno chousenminsyusyugijinminkyouwakoku [Preparations for Succession Begins: The Democratic People's Republic of Korea in 2010]," in *Ajia Doukou Nenpou 2011*

[Annual Report on Trends in Asia 2011] (Chiba, Japan: Institute of Developing Economies, 2011), 68; ROK Ministry of Unification, "bukhanjeongbopotal [North Korea Information Portal]".

21 ROK Ministry of Unification, "bukhanjeongbopotal [North Korea Information Portal]".

22 Ibid.; ROK Ministry of Unification, *bukhan gwonryeokgigudo [Organizational Chart of North Korean Leadership]* (Seoul, ROK: ROK Ministry of Unification, August 2016).

23 Rodong Sinmun, "joseonrodongdang jungangwiwonhoi je8gi je1chajeonwonhoieuie gwanhan gongbo [Press Release of First Plenary Meeting of Eighth WPK Central Committee]," *Rodong Sinmun*, 11 January 2021.

24 Yeong-hoon Lee, *bukhaneul umjikineun him: gunbueui paegwon gyeongjaeng [The Forces that Move North Korea: The Competition for Hegemony in the Military]* (Paju, ROK: Salim Books, 2012), 10–11.

25 Ibid., 27.

26 Scalapino and Lee, *Communism in Korea*, 934–35, 1381.

27 Institute of Developing Economies, "1972nenno kitachousen: jisyutekina syakaisyugikokkae [North Korea in 1972: Toward a Self-Reliant State]," in *Ajia Doukou Nenpou 1973 [Annual Report on Trends in Asia 1973]* (Chiba, Japan: Institute of Developing Economies, 1973), 99; ROK Ministry of Unification, "bukhanjeongbopotal [North Korea Information Portal]".

28 Motoi Tamaki, "kibishii 'koritsuka/keizaikonnan' dassyutsusakusen: 1990nenno chousenminsyusyugijinminkyouwakoku [The Challenging Escape from Isolation and Economic Difficulties: The Democratic People's Republic of Korea in 1990]," in *Ajia Doukou Nenpou 1991 [Annual Report on Trends in Asia 1991]* (Chiba, Japan: Institute of Developing Economies, 1991), 82; ROK Ministry of Unification, "bukhanjeongbopotal [North Korea Information Portal]".

29 Adrian Buzo, *Politics and Leadership in North Korea: The Guerilla Dynasty*, 2nd ed. (London, UK and New York: Routledge, 2018), 245.

30 Masahiko Nakagawa, "kwanmyonson1gouno uchiagede ishinkaifukuwo kokoromiru: 1998nenno chousenminsyusyugijinminkyouwakoku [Attempt at Recovering Legitimacy with the Launch of Kwangmyongsong-1: The Democratic People's Republic of Korea in 1998]," in *Ajia Doukou Nenpou 1999 [Annual Report on Trends in Asia 1999]* (Chiba, Japan: Institute of Developing Economies, 1999), 83; ROK Ministry of Unification, "bukhanjeongbopotal [North Korea Information Portal]".

31 Masahiko Nakagawa, "keizaikaikaku2kimeno naikakuseiritsu: 2003nenno chousenminsyusyugijinminkyouwakoku [Establishment of the Cabinet for the Second Term of Economic Reform: The Democratic People's Republic of Korea in 2003]," in *Ajia Doukou Nenpou 2004 [Annual Report on Trends in Asia 2004]* (Chiba, Japan: Institute of Developing Economies, 2004), 87; ROK Ministry of Unification, "bukhanjeongbopotal [North Korea Information Portal]".

32 Masahiko Nakagawa, "2domeno rokettohassyato kakujikken: 2009nenno chousenminsyusyugijinminkyouwakoku [The Second Rocket Launch and Nuclear Test: The Democratic People's Republic of Korea in 2009]," in *Ajia Doukou Nenpou 2010 [Annual Report on Trends in Asia 2010]* (Chiba, Japan: Institute of Developing Economies, 2010), 87.

33 ROK Ministry of Unification, "bukhanjeongbopotal [North Korea Information Portal]".

34 Ibid.

35 Ibid.

36 Ibid.

37 Ibid.

38 Ibid.

39 See: Michael Madden, "The Fourth Session of the 13th SPA: Tweaks at the Top," *38 North*, July 6, 2016, http://38north.org/2016/07/mmadden070616/.
40 Masahiko Nakagawa, "kakuheiki/misairu kaihatsuno shintento sono daisyou: 2016nenno chousenminsyusyugijinminkyouwakoku [Progress in Nuclear Weapons and Missile Developments and Its Costs: The Democratic People's Republic of Korea in 2016]," in *Ajia Doukou Nenpou 2017 [Annual Report on Trends in Asia 2017]* (Chiba, Japan: Institute of Developing Economies, 2017), 95; ROK Ministry of Unification, "bukhanjeongbopotal [North Korea Information Portal]".
41 ROK Ministry of Unification, "bukhanjeongbopotal [North Korea Information Portal]".
42 Ibid.
43 Beom-chul Shin and Jin-a Kim, "bukhanguneui cheje [The North Korean Military System]," in *bukhangun sikeurit ripoteu [North Korea Military Secret Report]*, ed. Yong-won Yoo, Beom-chul Shin, and Jin-a Kim (Seoul, ROK: Planet Media, 2013), 65.
44 Jae-hong Ko, *bukhangun choigosaryeonggwan wisang yeongu [Study on the Status of Supreme Commander of the North Korean Military]* (Seoul, ROK: Korea Institute for National Unification, 2006), 2.
45 For the most comprehensive description of the various MND bureaus and offices, see: Lee, *bukhaneul umjikineun him: gunbueui paegwon gyeongjaeng [The Forces that Move North Korea: The Competition for Hegemony in the Military]*, 59–66.
46 Ibid., 26–27.
47 See: Il-sung Kim, "inmingundaenae dangjeongchisaeobeul gaeseonganghwahagi wihan gwaeob [Tasks for Improving Party Political Work in the People's Army] (8 March 1958)," in *Kim Il Sung jeojakjib [Kim Il Sung Works]*, vol. 12 (Pyongyang, DPRK: Joseon Rodongdang Chulpansa, 1981).
48 See: Il-sung Kim, "inmingundaenaee joseonrodongdang dancheruel jojikhalde daehayeo [On Forming Workers' Party of Korea Organization in the People's Army] (21 October 1950)," in *Kim Il Sung jeojakjib [Kim Il Sung Works]*, vol. 6 (Pyongyang, DPRK: Joseon Rodongdang Chulpansa, 1981).
49 See: Il-sung Kim, "gunindeul sokeseo gongsanjueuigyoyanggwa hyeokmyeong jeontonggyoyangeul ganghwa halde daehayeo [On Strengthening the Education in Communism and in the Revolutionary Traditions Among Soldiers] (30 October 1958)," 560–80; Il-sung Kim, "inmingundaenae dangjeongchisaeobeseo gyojojueuireul bandaehago juchereul seulde daehayeo [On Opposing Dogmatism and Establishing Juche in Party Political Work in the People's Army] (16 May 1959)," in *Kim Il Sung jeojakjib [Kim Il Sung Works]*, vol. 13 (Pyongyang, DPRK: Joseon Rodongdang Chulpansa, 1981), 301–6; Il-sung Kim, "inmingundaenaeeseo jeongchisaeobeul ganghwahalde daehayeo [On Strengthening Political Work in the People's Army] (8 September 1960)," in *Kim Il Sung jeojakjib [Kim Il Sung Works]*, vol. 14 (Pyongyang, DPRK: Joseon Rodongdang Chulpansa, 1981).
50 See: Il-sung Kim, "inmingundaeeui dangjojiksaeobeul gaeseonhalde daehayeo [On Improving Party Organizational Work in the People's Army] (7 Nov 1969)," in *Kim Il Sung jeojakjib [Kim Il Sung Works]*, vol. 24 (Pyongyang, DPRK: Joseon Rodongdang Chulpansa, 1983).
51 See: Jong-il Kim, "songunhyeokmyeongroseoneun uri sidaeeui widaehan hyeokmyeongroseonimyeo uri hyeokmyeongeui baekjeonbaekseungeui gichiida [The Songun-Based Revolutionary Line is a Great Revolutionary Line of Our Era and an Ever-Victorious Banner of Our Revolution] (29 January 2003)," in *Kim Jong Il seonjib [Kim Jong Il Selected Works]*, vol. 21, Expanded ed. (Pyongyang, DPRK: Joseon Rodongdang Chulpansa, 2013), 234–35.

52 Scalapino and Lee, *Communism in Korea*, 963–65.
53 Dae-keun Yi, *bukhanguneun woae kudetareul haji ana [Why Don't the Korean People's Army Make a Coup]* (Paju, ROK: Hanul Academy, 2003), 176–80.
54 Ibid., 173–75.
55 ROK Ministry of Unification, *Understanding North Korea* (Seoul, ROK: ROK Ministry of Unification, 2017), 135.
56 Scalapino and Lee, *Communism in Korea*, 963.
57 Yi, *bukhanguneun woae kudetareul haji ana [Why Don't the Korean People's Army Make a Coup]*, 161–63.
58 See: Ibid., 161.
59 Lee, *bukhaneul umjikineun him: gunbueui paegwon gyeongjaeng [The Forces that Move North Korea: The Competition for Hegemony in the Military]*, 41.
60 Ibid., 37–38.
61 Joseph S. Bermudez Jr., *The Armed Forces of North Korea* (St. Leonards, Australia: Allen & Unwin, 2001), 177; Joseph S. Bermudez Jr., "38 North Special Report: A New Emphasis on Operations against South Korea?" *38 North*, June 11, 2010, www.38north.org/2010/06/a-new-emphasis-on-operations-against-south-korea/; Andrei N. Lankov, "On the Great Leader's Secret Service: North Korea's Intelligence Agencies," *NK News*, May 1, 2017, www.nknews.org/2017/05/on-the-great-leaders-secret-service-north-koreas-intelligence-agencies/.
62 The WPK still operates its own intelligence organs such as the United Front Department and the Liaison Department.
63 Bermudez Jr., "38 North Special Report: A New Emphasis on Operations Against South Korea?"; Lankov, "On the Great Leader's Secret Service: North Korea's Intelligence Agencies."
64 See: Lee, *bukhaneul umjikineun him: gunbueui paegwon gyeongjaeng [The Forces That Move North Korea: The Competition for Hegemony in the Military]*, 69.
65 NK Chosun, "buk, jaknyeon 4wol 'jeonsisaeobsechik' hadal [North Korea Issued the 'War Operation Guidelines' Last Year]," *NK Chosun*, January 5, 2005, http://nk.chosun.com/bbs/list.html?table=bbs_16&idxno=1943&page=41&total=2178&sc_area=&sc_word=.
66 Lee, *bukhaneul umjikineun him: gunbueui paegwon gyeongjaeng [The Forces that Move North Korea: The Competition for Hegemony in the Military]*, 16.
67 ROK Ministry of Unification, *bukhan gigwanbyeol inmyeongrok [Directory of North Korean People by Institutions]* (Seoul, ROK: Ministry of Unification, 2020), 282–91.
68 Lee, *bukhaneul umjikineun him: gunbueui paegwon gyeongjaeng [The Forces that Move North Korea: The Competition for Hegemony in the Military]*, 27.
69 Scalapino and Lee, *Communism in Korea*, 1001–2.
70 For example, Lee Yeong-hoon claimed that many in the GSD had little knowledge about the CCWPK secretariat. Lee, *bukhaneul umjikineun him: gunbueui paegwon gyeongjaeng [The Forces that Move North Korea: The Competition for Hegemony in the Military]*, 76.
71 Scalapino and Lee, *Communism in Korea*, 969–73; Taik-young Hamm, *Arming the Two Koreas: State, Capital and Military Power* (London, UK and New York: Routledge, 1999), 144. The Korean version of Kim Il-sung's speech found in: Lee, *Kim Jong-il chejeeui bukhangundae haebu [Anatomy of the Kim Jong-il Regime's North Korean Army]*, 48.
72 Lee, *bukhaneul umjikineun him: gunbueui paegwon gyeongjaeng [The Forces that Move North Korea: The Competition for Hegemony in the Military]*, 6, 74.
73 Kwang-soo Kim, "joseon inmingun changseolgwa baljeon [The Establisment and Development of the Korean People's Army]," in *bukhangunsamunjeeui jaejomyeong [The Military of North Korea: A New Look]* (Paju, ROK: Hanul Academy, 2006), 121.

74 Scalapino and Lee, *Communism in Korea*, 497–98.
75 For another discussion on the changes under Kim Jong-un, see: Michael Madden, "38 North Special Report: Recent Changes in Kim Jong Un's High Command," *38 North*, July 3, 2018, www.38north.org/2018/07/mmadden070318/.

References

Bermudez Jr., Joseph S. *The Armed Forces of North Korea*. St. Leonards, Australia: Allen & Unwin, 2001.

———. "38 North Special Report: A New Emphasis on Operations against South Korea?" *38 North*, June 11, 2010. www.38north.org/2010/06/a-new-emphasis-on-operations-against-south-korea/.

Buzo, Adrian. *Politics and Leadership in North Korea: The Guerilla Dynasty*. 2nd ed. London, UK and New York, NY: Routledge, 2018.

Choi, Wan-gyu. "joseoninminguneui hyeongseonggwa baljeon [The Formation and Development of the KPA]." In *bukhaneui gunsa [North Korean Military Affairs]*, edited by Bukhan Yeongu Hakhoi. Seoul, ROK: Gyeongin Munhwasa, 2006.

Hamm, Taik-young. *Arming the Two Koreas: State, Capital and Military Power*. London, UK and New York, NY: Routledge, 1999.

Institute of Developing Economies. "1972nenno kitachousen: jisyutekina syakaisyugikokkae [North Korea in 1972: Toward a Self-Reliant State]." In *Ajia Doukou Nenpou 1973 [Annual Report on Trends in Asia 1973]*. Chiba, Japan: Institute of Developing Economies, 1973.

Kim, Il-sung. "pyongyanghakwongaewonsikeul chukhahayeo [Congratulations on the Opening of the Pyongyang Institute] (23 February 1946)." In *Kim Il Sung jeojakjib [Kim Il Sung Works]*. Vol. 2. Pyongyang, DPRK: Joseon Rodongdang Chulpansa, 1980.

———. "gunindeul sokeseo gongsanjueuigyoyanggwa hyeokmyeong jeontonggyoyangeul ganghwa halde daehayeo [On Strengthening the Education in Communism and in the Revolutionary Traditions Among Soldiers] (30 October 1958)." In *Kim Il Sung jeojakjib [Kim Il Sung Works]*. Vol. 12. Pyongyang, DPRK: Joseon Rodongdang Chulpansa, 1981a.

———. "inmingundaenae dangjeongchisaeobeseo gyojojueuireul bandaehago juchereul seulde daehayeo [On Opposing Dogmatism and Establishing Juche in Party Political Work in the People's Army] (16 May 1959)." In *Kim Il Sung jeojakjib [Kim Il Sung Works]*. Vol. 13. Pyongyang, DPRK: Joseon Rodongdang Chulpansa, 1981b.

———. "inmingundaenae dangjeongchisaeobeul gaeseonganghwahagi wihan gwaeob [Tasks for Improving Party Political Work in the People's Army] (8 March 1958)." In *Kim Il Sung jeojakjib [Kim Il Sung Works]*. Vol. 12. Pyongyang, DPRK: Joseon Rodongdang Chulpansa, 1981c.

———. "inmingundaenaee joseonrodongdang danchereul jojikhalde daehayeo [On Forming Workers' Party of Korea Organization in the People's Army] (21 October 1950)." In *Kim Il Sung jeojakjib [Kim Il Sung Works]*. Vol. 6. Pyongyang, DPRK: Joseon Rodongdang Chulpansa, 1981d.

———. "inmingundaenaeeseo jeongchisaeobeul ganghwahalde daehayeo [On Strengthening Political Work in the People's Army] (8 September 1960)." In *Kim Il Sung jeojakjib [Kim Il Sung Works]*. Vol. 14. Pyongyang, DPRK: Joseon Rodongdang Chulpansa, 1981e.

———. "inmingundaeeui dangjojiksaeobeul gaeseonhalde daehayeo [On Improving Party Organizational Work in the People's Army] (7 Nov 1969)." In *Kim Il Sung jeojakjib [Kim Il Sung Works]*. Vol. 24. Pyongyang, DPRK: Joseon Rodongdang Chulpansa, 1983.

Kim, Jong-il. "songunhyeokmyeongroseoneun uri sidaeeui widaehan hyeokmyeongroseonimyeo uri hyeokmyeongeui baekjeonbaekseungeui gichiida [The Songun-Based Revolutionary Line is a Great Revolutionary Line of Our Era and an Ever-Victorious Banner of Our Revolution] (29 January 2003)." In *Kim Jong Il seonjib [Kim Jong Il Selected Works]*. Vol. 21. Expanded ed. Pyongyang, DPRK: Joseon Rodongdang Chulpansa, 2013.

Kim, Kwang-soo. "joseon inmingun changseolgwa baljeon [The Establishment and Development of the Korean People's Army]." In *bukhangunsamunjeeui jaejomyeong [The Military of North Korea: A New Look]*. Paju, ROK: Hanul Academy, 2006.

Ko, Jae-hong. *bukhangun choigosaryeonggwan wisang yeongu [Study on the Status of Supreme Commander of the North Korean Military]*. Seoul, ROK: Korea Institute for National Unificationa, 2006.

Komaki, Teruo. "roudoutoudairokkaitaikaino toshi: 1980nenno chousenminsyusyugijinminkyouwakoku [The Year of the Sixth Congress of the Workers' Party of Korea: The Democratic People's Republic of Korea in 1980]." In *Ajia Doukou Nenpou 1981 [Annual Report on Trends in Asia 1981]*. Chiba, Japan: Institute of Developing Economies, 1981.

———. "'kunanno kougun' kade kimujonirutaiseigaseisikihossoku: 1997nenno chousenminsyusyugijinminkyouwakoku [The Official Inauguration of the Kim Jong-il Regime under the 'Arduous March': The Democratic People's Republic of Korea in 1997]." In *Ajia Doukou Nenpou 1998 [Annual Report on Trends in Asia 1998]*. Chiba, Japan: Institute of Developing Economies, 1998.

Lankov, Andrei N. "On the Great Leader's Secret Service: North Korea's Intelligence Agencies." *NK News*, May 1, 2017. www.nknews.org/2017/05/on-the-great-leaders-secret-service-north-koreas-intelligence-agencies/.

Lee, Min-ryong. *Kim Jong-il chejeeui bukhangundae haebu [Anatomy of the Kim Jong-il Regime's North Korean Army]*. Seoul, ROK: Hwanggeumal, 2004.

Lee, Yeong-hoon. *bukhaneul umjikineun him: gunbueui paegwon gyeongjaeng [The Forces that Move North Korea: The Competition for Hegemony in the Military]*. Paju, ROK: Salim Books, 2012.

Madden, Michael. "38 North Special Report: Recent Changes in Kim Jong Un's High Command." *38 North*, July 3, 2018. www.38north.org/2018/07/mmadden070318/.

———. "The Fourth Session of the 13th SPA: Tweaks at the Top." *38 North*, July 6, 2016. http://38north.org/2016/07/mmadden070616/.

Nakagawa, Masahiko. "'idaina syuryou' no shikyo: 1994nenno chousenminsyusyugijinminkyouwakoku ['The Death of the Great Suryong': The Democratic People's Republic of Korea in 1994]." In *Ajia Doukou Nenpou 1995 [Annual Report on Trends in Asia 1995]*. Chiba, Japan: Institute of Developing Economies, 1995.

———. "kwanmyonson1gouno uchiagede ishinkaifukuwo kokoromiru: 1998nenno chousenminsyusyugijinminkyouwakoku [Attempt at Recovering Legitimacy with the Launch of Kwangmyongsong-1: The Democratic People's Republic of Korea in 1998]." In *Ajia Doukou Nenpou 1999 [Annual Report on Trends in Asia 1999]*. Chiba, Japan: Institute of Developing Economies, 1999.

———. "keizaikaikaku2kimeno naikakuseiritsu: 2003nenno chousenminsyusyugijinminkyouwakoku [Establishment of the Cabinet for the Second Term of Economic

Reform: The Democratic People's Republic of Korea in 2003]." In *Ajia Doukou Nenpou 2004 [Annual Report on Trends in Asia 2004]*. Chiba, Japan: Institute of Developing Economies, 2004.

———. "2domeno rokettohassyato kakujikken: 2009nenno chousenminsyusyugijinminkyouwakoku [The Second Rocket Launch and Nuclear Test: The Democratic People's Republic of Korea in 2009]." In *Ajia Doukou Nenpou 2010 [Annual Report on Trends in Ajia 2010]*. Chiba, Japan: Institute of Developing Economies, 2010.

———. "koukeitaiseikouchikuno junbi hajimaru: 2010nenno chousenminsyusyugijinminkyouwakoku [Preparations for Succession Begins: The Democratic People's Republic of Korea in 2010]." In *Ajia Doukou Nenpou 2011 [Annual Report on Trends in Asia 2011]*. Chiba, Japan: Institute of Developing Economies, 2011.

———. "kakuheiki/misairu kaihatsuno shintento sono daisyou: 2016nenno chousenminsyusyugijinminkyouwakoku [Progress in Nuclear Weapons and Missile Developments and its Costs: The Democratic People's Republic of Korea in 2016]." In *Ajia Doukou Nenpou 2017 [Annual Report on Trends in Asia 2017]*. Chiba, Japan: Institute of Developing Economies, 2017.

NK Chosun. "buk, jaknyeon 4wol 'jeonsisaeobsechik' hadal [North Korea Issued the 'War Operation Guidelines' Last Year]." *NK Chosun*, January 5, 2005. http://nk.chosun.com/bbs/list.html?table=bbs_16&idxno=1943&page=41&total=2178&sc_area=&sc_word=.

Rodong Sinmun, "joseonrodongdang jungangwiwonhoi je8gi je1chajeonwonhoieuie gwanhan gongbo [Press Release of First Plenary Meeting of Eighth WPK Central Committee]," *Rodong Sinmun*, January 11, 2021.

ROK Ministry of Unification. *bukhan gigwanbyeol inmyeongrok [Directory of North Korean People by Institutions]*. Seoul, ROK: Ministry of Unification, 2020.

———. *bukhan gwonryeokgigudo [Organizational Chart of North Korean Leadership]*. Seoul, ROK: ROK Ministry of Unification, August 2016.

———. "bukhanjeongbopotal [North Korea Information Portal]." nkinfo.unikorea.go.kr.

———. *Understanding North Korea*. Seoul, ROK: ROK Ministry of Unification, 2017.

Scalapino, Robert A., and Chong-sik Lee. *Communism in Korea*. Berkeley, CA: University of California Press, 1972.

Shin, Beom-chul, and Jin-a Kim. "bukhanguneui cheje [The North Korean Military System]." In *bukhangun sikeurit ripoteu [North Korea Military Secret Report]*, edited by Yong-won Yoo, Beom-chul Shin and Jin-a Kim. Seoul, ROK: Planet Media, 2013.

Tamaki, Motoi. "kibishii 'koritsuka/keizaikonnan' dassyutsusakusen: 1990nenno chousenminsyusyugijinminkyouwakoku [The Challenging Escape from Isolation and Economic Difficulties: The Democratic People's Republic of Korea in 1990]." In *Ajia Doukou Nenpou 1991 [Annual Report on Trends in Asia 1991]*. Chiba, Japan: Institute of Developing Economies, 1991.

Tsukamoto, Katsuichi. *kitachousengunto seiji [The North Korean Army and Politics]*. Tokyo, Japan: Hara Shobo, 2000.

Yi, Dae-keun. *bukhanguneun woae kudetareul haji ana [Why Don't the Korean People's Army Make a Coup]*. Paju, ROK: Hanul Academy, 2003.

4 Economic and industrial capacity

For any state, the economy is pivotal in defense planning, where costs and benefits frame the thinking of decision-makers. While it is the latter that allures decision-makers, in almost all cases it is the former that constrains their pursuit of the best options available. Naturally, the capacity to finance the military sector much depends on the state's economic and industrial capacity. Greater challenges are faced when states pursue autonomy by building their own military-industrial complex – particularly when they are economically and technologically constrained in their capacity. The DPRK presents an interesting case, where they have pursued to establish an indigenous military industry for two reasons. First, much was based on not only the nationalistic vision that an autonomous military industry would be a benchmark as an independent state but also the idea that the focus on heavy industry would kill two birds with one stone by strengthening the armed forces while modernizing the economy. Second, the DPRK felt a great deal of vulnerability, and questioned the credibility and reliability of its allies in providing Pyongyang with the necessary technologies to strengthen the KPA. Under the auspices of Juche, the DPRK has taken significant steps to establish its own military–industrial complex. Yet despite the commitment, there were significant problems in both capacity and policies that led to mixed results.

The development of the military-industrial complex

Since the state's formative years, Kim Il-sung envisioned to establish an indigenous military-industrial complex in the DPRK. In October 1949, Kim Il-sung argued that although it is cheaper to purchase weapons from abroad, doing so was unsafe from the national security standpoint as it would up the DPRK's dependence on other states without solid guarantees.[1] To a great extent, North Korea's industrialization drew on the industrial infrastructures left behind by the Japanese, as well as the material and technical aid from the USSR. Still, the actual progress in building the DPRK's autonomous military–industrial capacity was incremental. First, the Japanese dismantled or destroyed many of the key industrial facilities when their colonial rule was over, requiring much effort to bring them back to operation. Second, the Soviets had sabotaged and scavenged the various industrial facilities left behind by the Japanese for their own material gains. Third, North Korea lacked raw materials and skilled workers, constraining the immediate operation of industrial assets. Fourth, the economic plans of the first several years also focused much on nationalizing major industries and executing land reforms, making the establishment of the military industry one of the many agendas Pyongyang was pursuing. Fifth, while there were many factories in the northern half of the Korean peninsula built by the Japanese for their war effort, much of the industrial infrastructures were for materials to feed the war effort such as mining of coals and minerals, production of chemicals, metals, and electricity. The only significant weapons factory North Korea inherited from the Japanese was the Pyongyang Munitions Manufacturing Plant that was renamed as Factory No. 25 (later renamed as Factory No. 65 and now the February 8 Machine Complex) specialized in small arms and ammunitions.[2] Thus, the developments in the late 1940s were more about the establishment of the industrial-base rather than capacity to arm the KPA.

The Korean War naturally increased the demands for weapons and supplies. Still, the DPRK was much dependent on the weapons brought in from the USSR, and its autonomous production capacity was limited to small arms. Moreover, the North Korean war factories were constructed under a wartime environment, making them vulnerable to attacks as well as suffering from disruptions in the logistical supply chain. Number of factories were dispersed and relocated to dodge the UN forces. For example, Factory No. 65 was divided and dispersed into a number of munitions plants and also constructed new ones, such as Factory No. 26, Factory No. 42, Factory No. 76, Factory No. 107, Factory No. 145, and Factory No. 205.[3] Moreover, Kim Il-sung also stressed the importance of constructing the key chemical, machine, ordnance, and steel factories underground.[4]

The developments in the DPRK military industry during the immediate post-armistice period were slowed by the war-torn economy and society, forcing Pyongyang to devote much of its resources to state economic reconstruction under the nine-year Three-Stage Plan for Post-War Reconstruction. In 1953 and 1954, the DPRK received a significant amount of material and monetary aid from the USSR, China, and the selected Eastern European states that significantly fueled Pyongyang's rapid economic reconstruction.[5] Moreover, the period between the mid-1950s and the early 1960s was also when Kim Il-sung was molding the economic system under greater party control. The same approach was applied to the military–industrial sector, where on 28 May 1961, Kim Il-sung gave a speech to WPK officials in the ordnance industry arguing that the military industry sector must develop through mass mobilization and strict ideological loyalty to the WPK.[6] The politicized measures certainly did not substantively boost the actual capacity of the military industry sector, but they did help Kim Il-sung gain the means to mobilize the industrial sector to work in accordance with the WPK lines.

The demands to build the indigenous military-industrial complex accelerated in the 1960s as the DPRK introduced the Line of Self-Reliant Defence. Kim Il-sung's fears about the vulnerabilities of dependence on its benefactors were fast becoming a reality, particularly as the DPRK witnessed the Sino–Soviet rift that led to the USSR's abrupt withdrawal of technical assistance to China. Pyongyang feared it too could suffer the same fate should relations with either or both Beijing and Moscow sour. While the treaties signed with both China and the USSR in 1961 provided various military assets for the short term, the DPRK sought to attain greater autonomous capacity and means of producing military hardware. In his speech on the Three Revolutionary Forces for Reunification, Kim Il-sung specifically emphasized the importance of autonomously producing modern weapons as a critical part of strengthening the KPA.[7] At the Fifth WPK Congress in November 1970, Kim Il-sung claimed that the industrial foundations to achieve the Line of Self-Reliant Defence were now in place.[8] Still, the developments during the 1960s were primarily about establishing the industrial foundations by constructing the relevant factories and logistical supply chains. Hence, although the DPRK was steadily building its industrial capacity to produce arms, it was not yet ready to fully self-reliantly produce modern systems for the KPA.[9]

The DPRK also worked to establish a governance mechanism to exercise direct, centralized command and control over the factories, R&D facilities, and supply chains. After the armistice, the weapons factories were being shifted around various ministries and bureaus, including the Ministry of Heavy Industries, Ministry of Light Industries, First Bureau of the Cabinet, and then the First Bureau of the Ministry of Machine Industry.[10] As the economy came under greater centralized control, the military industry was placed under a more robust regime. In the early 1960s, Pyongyang established the Second Machinery Industry Department subordinate to the Cabinet. However, the growth of the DPRK's military industry sector combined with Kim Il-sung's continued push for centralization and politicization required further administrative reconfigurations. As a result, the Second Machinery Industry Department was reorganized as the Second Economic Commission (SEC) in 1971 that works under the MID of the CCWPK.[11] Despite the limitations in capacity, the establishment of the SEC undoubtedly marked a new stage in the DPRK's quest to establish its own military-industrial complex for the autonomous production of military hardware.

Although the DPRK provides virtually no information about the SEC, there is a general consensus that it consists of the: General Planning Bureau, seven general bureaus, one trade bureau as well as R&D institutions. At the top, the General Planning Bureau handles the overall administrative matters such as budgeting, planning as well as acquisition and distribution of necessary resources. The seven numbered general bureaus oversee the 200 or so ordnance factories as well as an unknown but large number of civilian factories that produce weapons, equipment, and other material as well as conducting major repairs.[12] Each of the numbered bureaus focuses on particular systems as given below:[13]

- First General Bureau: Small arms, ammunitions
- Second General Bureau: Tanks, armored vehicles, towing vehicles
- Third General Bureau: Artillery, anti-air guns, MLRS, mortars
- Fourth General Bureau: Anti-air/ship/tank missiles, ballistic missiles

- Fifth General Bureau: Nuclear, chemical, biological weapons
- Sixth General Bureau: Surface vessels, submarines, hovercraft
- Seventh General Bureau: Communications equipment, aircraft

The Academy of the National Defence Science (formerly known as the Second Academy of Natural Sciences) is in charge of not only conducting indigenous R&D of science and technology for military purposes but also to study imported systems.[14] According to Kwon, the Academy of the National Defence Science has over 30 research centers relating to chemistry, electronics, engineering, information and communication technology (ICT), physics, metallurgy, shipbuilding, and others.[15] Although it is widely believed that the Academy of the National Defence Science is the primary institution in the R&D of military-related science and technology, it is likely that other institutions are also involved, including various universities and research institutions specializing in science and technology that have the capacity to provide know-how and produce "spin-on" technologies for the military sector. For example, both the Atomic Energy Institute and the National Aerospace Development Administration have been widely believed to have played important roles in Pyongyang's development of nuclear weapons and ballistic missiles, respectively.

The External Economic Bureau is in charge of not only buying and selling weapons and related technologies but also possibly other commercial activities to finance the military sector. One of the primary companies affiliated to the External Economic Bureau is the Yongaksan Trading Company, as well as other companies that are subordinate to, or coordinates with the bureau for foreign procurements and sales (e.g., companies belonging to the MND). In addition, the External Economic Bureau is known to be running its own financial institutions such as the Kumgang Bank to handle the transactions by the bureau.[16]

Despite its enormous and sophisticated structure, the SEC functions as an acquisition and production management system rather than being a decision-making organ. Just like the KPA, the SEC is governed by a multi-dimensional structure under the WPK and the SAC to ensure that acquisitions and production are consistent with the leadership's defense planning decisions and directions. On the party side, Article 27 of the WPK charter clearly states that the CMCWPK has control over affairs relating to the military industry. However, although the CMCWPK issues the orders, the SEC is handled by the MID of the CCWPK. On the state side, the military industry has worked under the CPC since 1972, and then the NDC since 1998, making it likely that the SAC has inherited the hierarchy by having executive control over the military-industrial complex.[17]

There are questions concerning the relationship between the SEC and the relevant ministries. Logically, the SEC would liaise with the MND concerning demands, R&D, procurements, and delivery.[18] Yet given that the SEC is about industrial coordination and oversight of factories, it is apt to be connected to the various industry-related ministries such as not only the Ministry of Machine-Building Industry but also the Ministry of Chemical Industry, Ministry of Construction and Building-Materials Industry, Ministry of Consumer Goods

Industry, Ministry of Electric Power Industry, Ministry of Electronics Industry, Ministry of External Economic Relations, Ministry of Land and Maritime Transport, Ministry of Light Industry, Ministry of Metallurgical Industry, Ministry of Mining Industry, Ministry of Posts and Telecommunications, Ministry of Railways, Ministry of State Construction Control, Ministry of State Natural Resources Development, State Academy of Sciences, and the State Science and Technology Commission. While there is little doubt about the linkage between the SEC and the ministries that facilitate the acquisitions and production for the military sector, the processes are not independent and can only take place with the reviews and approvals from the CMCWPK and SAC.

Information about former and incumbent cadres in the MID and SEC have been intermittent at best, and only a limited number of names are decipherable, such as Han Song-ryong, Hong Sung-mu, Hong Yong-chil, Jo Chun-ryong, Jon Pyong-ho, Ju Kyu-chang, Kim Chol-man, O Su-yong, Paek Se-bong, Pak Song-bong, Pak To-chun, Ri Man-gon, Ri Pyong-chol, and Thae Jong-su. Generally, the directors and deputies of the MID and SEC are either WPK technocrats or retired KPA officers who have spent a significant amount of their careers in the military industry. In most cases, the director of the MID has served in the CMCWPK and the NDC/SAC, as well as other senior positions in the CCWPK and the Political Bureau of the CCWPK. Often, the senior MID and SEC cadres are identified after collating the information of senior cadres in the CMCWPK, NDC/SAC, and CCWPK with those accompanying the leader in key weapons tests and guidance tours of military factories. In recent decades, Jon Pyong-ho, Ju Kyu-chang, Kim Chol-man, Paek Se-bong, and Pak To-chun have played particularly important roles in the development of nuclear weapons and ballistic missiles. Currently, Ri Pyong-chol, a former commander of the KPAAF, is reported to have be the current director of the MID.[19]

Assessing the DPRK's economic and industrial capacity

The establishment of the autonomous military-industrial complex was central to the DPRK's quest to achieve the Line of Self-Reliant Defence. Yet what seemed revolutionary in theory proved to be cost-inefficient and at times disastrous in practice, particularly as Pyongyang pursued its plans through centralized and politicized methods despite its limited resource and technological capacity. The problematic approaches inevitably led to bleak plans-reality mismatches that consequently undermined Pyongyang's autonomous ability to enhance the KPA's readiness.

Resource capacity

The Line of Self-Reliant Defence and the very size of the military-industrial complex required the DPRK to devote a significant amount of resources to the military sector. Yet analyzing the DPRK's military outlays is extremely troubling due to the lack of credible data on the precise amount and the breakdown of the expenditures such as resource allocation (e.g., personnel, hardware/infrastructures, O&M, R&D, etc.). Although military expenditures of socialist states have always been difficult to analyze, the DPRK's level of opaqueness far exceeds those of even China and the USSR, making the topic a black box even for the most seasoned experts.

Every year, the DPRK reports its national budget set by the State Planning Commission to the SPA. In most of the cases, the military expenditures are also announced, based on the percentage of the state expenditures allocated to the military sector. Yet the DPRK's official military outlays have been far from consistent since the dramatic increase in the 1960s. The DPRK initially concealed its military outlays between 1962 and 1966, but at the Fifth WPK Congress in November 1970, the then First Deputy Premier Kim Il reported that Pyongyang had spent a total of almost DPRK Won (KPW) 8.9 billion between 1961 and 1969.[20] The military outlays recorded 30.4% in 1967, and bounced between 31% and 32.4% from 1968 until 1971.[21] Hamm calculated that the DPRK's official military expenditures (in percentage of total government expenditures) averaged 19.8% between 1961 and 1966, and then 30.9% between 1967 and 1971.[22] Then since 1972, the official military expenditures dropped to 17.0% of total government expenditures and have since oscillated between 11.4% and 16.7%.

Outside analyses have often disputed the DPRK's official statistics, arguing that Pyongyang conceals or water-downs its military spending. Analyses conducted by the ROK and US governments, thinktanks such as the International Institute for Strategic Studies and the Stockholm International Peace Research Institute, as well as a number of scholars have often claimed that the actual percentage share of the DPRK state expenditures devoted to the military sector are far above Pyongyang's official figures. For example, the US State Department calculated that the DPRK spent approximately 23.3% of its GDP on the military between 2007 and 2017.[23] In 2016, the then ROK Minister of National Defense Han Min-koo claimed North Korea's military spending in 2013 was USD 10 billion.[24] Yet pinpointing exactly how much the DPRK spends on the military sector has proved to be notoriously difficult due to the lack of credible statistics and information about the state's accounting systems, but also issues concerning exchange rates.[25] Consequently, such problems have led to various methodologies and hypotheses over what *could* be the DPRK's real military expenditures.

One major issue in understanding the DPRK's military spending stems from the lack of clarity on what constitutes military outlays. According to the North Korean Kwangmyong Encyclopedia, military expenditures are defined as costs to finance the Line of Self-Reliant Defence doctrine.[26] Yet there is much ambiguity in the DPRK's description considering the broad nature of its defense planning doctrine that includes a variety of not only material components such as armaments, infrastructures, and human resources but also less-material aspects such as the "cadre army" guideline. Therefore, the description provided by the Kwangmyong Encyclopedia and other DPRK sources says little more than the obvious fact that its military spending purports to comprehensively strengthen the KPA without elaborating on the allocation of the resources.

In some respects, the DPRK's military spending patterns do make sense. The extremely high amount of military outlays in the 1960s coincides with the First Seven Year Plan that was configured to finance the Line of Self-Reliant Defence doctrine, involving the establishment and expansion of the military–industrial complex, construction of underground infrastructures, acquisition of platforms and weapons systems, and resources for wartime mobilization. Naturally, such projects were extremely expensive – particularly as Pyongyang was in haste to achieve those objectives. Nevertheless, the major problem is making sense of the military outlays since 1972, where the patterns clearly contradict with the developments in the KPA. Despite the sharp economic decline from the 1970s that plummeted into negative growth in the 1990s, the percentage of state expenditures devoted to the military sector since 1972 remained relatively consistent in comparison, suggesting that Pyongyang's military outlays would have decreased. Yet it was precisely the period between the 1970s and 2000s when the KPA acquired a range of conventional and strategic weapons, as well as boosting the number of active personnel.

The contradictions suggest that the DPRK may not have simply fudged the numbers, but in fact changed the definition of military outlays. Scholars who analyze the DPRK's military expenditures broadly agree that Pyongyang's official military outlays since 1972 only include O&M costs but exclude investment costs for R&D and procurements – a form of categorization that was also employed by China and the USSR.[27] The question, then, is how the investments are financed. Excluding the hardware received from China and the USSR in the form of aid and concessions, the DPRK would have devoted large sums of resources for the procurement and R&D of military systems. If the drop in military outlays since 1972 is due to changes in the categorization that only includes O&M costs, then the investment costs for procurement and R&D would be financed under a different classification. For example, Sung et al. argued that the costs for procurements and R&D related to the "Second Economy" are covered by "People's Economic Expenditures."[28] The argument does make sense, particularly if we take the idea that military outlays are those devoted to the MND who largely takes care of logistical matters while the R&D is tasked by the SEC that is separate from the MND.

If the DPRK's official military expenditures are indeed based purely on O&M, then there is some level of consistency, as the decrease in claimed outlays would explain the poor state of platforms and facilities, shortage of supplies, and welfare of soldiers. In readiness, the chronic shortages of fuel have compromised the quality of training and the shortage of spare parts have also undermined the state of KPA inventory. Furthermore, the KPA units have often been mobilized for agricultural, commercial, and industrial activities to fund their own welfare, much like how local communities are left to be self-reliant and self-sufficient. Indeed, this is a common, long-standing practice in many socialist states to offset the burdens in managing the units at the micro levels while focusing on macro agendas relating to the armed forces. Yet the DPRK's failure to adequately provide the logistical essentials to feed the massive KPA clearly evidence how the O&M outlays have been insufficient.

It is also important to note that the official military outlays are based on state expenditures which is only a portion of the resources received and used by the military sector. The State Planning Committee Military Planning Bureau staffed by approximately 150 KPA officers is in charge of allocating resources to the military and related industries but is essentially controlled by the CMCWPK.[29] The special status would indicate that funds and other resources relating to the military are, in fact, processed under separate regimes. Based on this, it is possible that investments for R&D and procurements are covered by extra-budgetary resources that are not reflected in the official outlays (e.g. MND and WPK's commercial activities and others).[30] In such case, the DPRK's official military outlays may not be completely fabricated, where the state genuinely provided the claimed amount to the MND for O&M while the WPK and MND generates funds to finance (or supplement) the costs for the acquisition of platforms and possibly fill some of the O&M shortages.

The WPK's commercial activities are complex, with the CCWPK Finance and Accounting Department Bureau 39 operating a number of companies known as "General Bureaus" such as Bonghwa, Daehung, Daesong, Kumgang, Kyonghung, and Rakwon that consist of various commercial, financial, industrial, and trading affiliates.[31] The WPK's commercial activities are diverse, ranging from the export of goods and services listed in not only official trade (e.g., agriculture and fisheries products, machinery, mining, processed foods, textiles, tourism, and transport) but also illicit trade of contraband products, counterfeit currency, financial fraud, hacking, narcotics, and others.

The MND also conducts its own commercial activities since the 1960s to shore up revenue. In 1995, the DPRK established an office within the ministry known as Department 44 that collectively manages the military's commercial activities and companies.[32] The MND owns commercial groups such as Maebong, Moran, Ryungsong groups and also runs its own banks to handle the finances and transactions.[33] Some of the MND's companies are based overseas, with the recent case being Glocom based in Malaysia that sold military communications and radio equipment. Naturally, the key commercial activities by the MND are arms trade and consulting (e.g., construction/production, trade, training, etc.) but is also widely believed to include non-military commercial activities such as labor and trading of raw materials and processed goods, as well as various illicit activities.

Official North Korean information on the WPK and MND's commercial activities are minimal. Even regarding the companies listed by the UN and state governments as target of sanctions, almost all of them are front companies that periodically change their identities to dodge foreign intelligence and international sanctions. Hence, details on the funds generated and provided by the WPK and MND enterprises are unknown. Furthermore, another caveat concerning the WPK and MND's commercial activities is that not all the earnings are transferred to the military sector as a certain portion would be channeled to the leadership in the form of "royalty tributes."[34]

While the funds generated by the WPK and MND are nonetheless vital to the DPRK, there are questions concerning exactly how much is channeled to the KPA's development and whether the income has been stable – particularly in regards to arms trade. In general, the DPRK's arms export market is limited to countries or non-state actors that are either economically or legally restricted from weapons acquisitions, making Pyongyang the last resort. Some of the countries known to have imported the DPRK's military products include Cuba, the Democratic Republic of Congo, Egypt, Ethiopia, Iran, Iraq (pre-2003), Libya, Myanmar, Pakistan, Syria, Vietnam, Yemen, and Zimbabwe.[35] Although the actual revenue earned by the DPRK from arms trade is unknown, it was estimated that Pyongyang earned as much as USD 100 million per year through the export of weapons systems and components.[36] Pyongyang's arms trade has been far from stable, with some claiming that its arms trade reached its peak in the 1980s but then dropped sharply in the 1990s.[37] Moreover, the DPRK's commercial activities have faced strong headwinds since the mid-2000s with the successive UN Security Council resolutions that target a variety of entities involved in the trading of military-related technologies, or commercial activities suspected to be linked to the military sector.[38] Pyongyang's illicit trade activities have been probed and seized by an increasing number of countries, compelling the DPRK to trade military technologies through more indirect means, such as via third-party contractors and other complex routes. Indeed, there is no evidence to suggest that the DPRK's arms trade has stopped as much would be going on behind the scenes. Even in recent years, the DPRK's arms trade has continued in illicit forms, most notably with Iran and Syria.[39] Yet still, mysteries remain as to exactly how much arms trade is stably financing the North Korean military sector when one considers the up-and-downs in arms trade and the high military expenditures.

Returning to the question about whether the DPRK spends enough on the KPA, much comes down to how the DPRK allocates its limited resources. While Pyongyang undoubtedly devotes a large portion of its economy to the military sector in percentage terms, the absolute amount is small given the size of the economy. Even with the higher-end estimates that the DPRK spends USD 10 billion (or more) on the military, that amount is still less than a quarter of ROK's defense spending and one-fifth of Japan's.[40] Indeed, the per capita costs for salaries, supplies, and maintenance of infrastructures and platforms in the DPRK would be lower compared to other states. Nevertheless, when one considers the actual size of the KPA and the military-industrial complex, the costs to strengthen and sustain the military sector would still be significantly high. Consequently, there would be dilemmas in the allocation of resources for investments and O&M. Keeping old equipment may save investment costs but ups the O&M costs to keep the hardware operational, forcing trade-offs with other items such as procurements, salaries, and so forth. Hence, the gap between the limited amount of resources available and the very size of the military sector ups the stakes for Pyongyang to strengthen the KPA, where poor budgeting and allocation would actually make the military expenditures insufficient.

Technological capacity

Since the formative years, the DPRK has emphasized the importance of science and technology for state development. The focus on science and technology is written in the constitution, vowing to enhance the DPRK's ability to build its autonomous industrial capacity. While there is much political rhetoric to justify and glorify the Juche-ist revolution guided by the leadership, the DPRK has genuinely undertaken various multi-year plans for developments in science and technology.

The DPRK has been blessed with a relatively favorable resource base to produce industrial materials. During the colonial era, the key reason for Japan's industrialization of northern part of the Korean peninsula was not simply due to the proximity to China, but because of the abundance of resources such as minerals to produce materials for the heavy chemical industry. The DPRK inherited this logic in the post-liberation period but through the centralized and mass-mobilized Stalinist approach, pushing to industrialize with focus on chemicals, metals, and machine-building. Up until the 1960s, the DPRK built a relatively strong capacity to transform raw materials into industrial materials and products – albeit with the exception of oil and gas where Pyongyang remained dependent on its benefactors. Thus, although the scale of the DPRK's industrial power is certainly not large, Pyongyang did become relatively self-sufficient in producing some materials such as concrete and metals that are essential to produce goods, machines, and infrastructures for the heavy industry sector.

In human resources, despite the initial quantitative shortage of skilled personnel, much was turned around in the first couple of decades after liberation. Soon after liberation, a number of scientists such as Ri Sung-gi, Ryo Kyong-ku, and To Sang-rok educated by the Japanese joined the DPRK and later played significant roles in the science and technology fields – particularly in chemistry and engineering. Moreover, the DPRK since the early years has focused much on educating and training technocrats to qualitatively and quantitatively enable the DPRK's indigenous R&D capacity. The period of mandatory education in the DPRK is 12-years, and while the schools are known for their intense ideological education, they also run curriculums that emphasize science and technology. As for the tertiary education system, much is modeled on the USSR, with a high ratio of specialized universities and colleges focusing on science and technology. The oldest, and most prestigious is the Kim Chaek University of Technology, but there are also many other higher education institutions focusing on science and technology such as the Chongjin Mine and Metal University, Chongjin University of Technology, Hamhung University of Chemical Industry, Hamhung University of Mathematical and Physical Sciences, Huichon University of Telecommunications, Pyongyang University of Construction and Building Materials, Pyongyang University of Machinery, Pyongyang University of Mechanical Engineering, Pyongyang Medical University, Pyongyang University of Science and Technology, Rajin University of Marine Transport, Sinuiju University of Light Industry, State Academy of Sciences, and others.

Despite the efforts for industrialization, the DPRK's production capacity has been limited and frequently disrupted. Above all, the DPRK has been historically constrained from stable access to energy, where the rate and volume of industrialization outpaced the energy supplies. While there are a number of oil refineries and chemical plants, the DPRK remains dependent on imported oil and gas. In oil, Moscow was the main supplier during the Cold War, but then faced a sharp decrease in the supplies since the collapse of the USSR. In the post-Cold War era, China became the biggest supplier of various fuels, although the actual amount has been far from lucrative and sufficient, and further worsened due to international sanctions. Even regarding electricity, the DPRK does have an array of coal and hydroelectric power plants that generate almost all of its electricity. Yet much of the infrastructures are those that were left behind by Japanese or built with the aid from socialist states during the Cold War, making them insufficient in meeting the DPRK's demands. Thus, although the military sector may be prioritized, the energy shortages have been so chronic that they nonetheless affect the ordnance factories' production as well as the logistical supply chain.

While one could argue that the DPRK could expand and upgrade its energy infrastructures and diversify its international network to procure oil and gas, doing so is difficult in practice, not simply due to the immense costs of contracting deals and installing power infrastructures but also the current international sanctions. Nuclear energy has been pursued as one of the solutions particularly with the construction of light water reactors, yet the actual progress is currently unknown. Moreover, Kim Jong-un is also pursuing renewable energy, with the plans handled by the Natural Energy Research Centre to construct solar and wind power stations. Yet although the DPRK's steps in renewable energy have its share of positives, it is premature to argue that they will lead to an immediate turnaround in the state's industrial capacity and are certainly not substitutes for oil and gas that remain to be essential for the heavy industry and military sectors.

Another endemic problem is the outdated nature of the industrial infrastructures that constrains the rate and quality of production. Indeed, the DPRK industrialized rapidly during the early decades, and there were also some credible developments since Juche-nization. Yet as time progressed, the DPRK's industrial infrastructures and machinery increasingly began to trail far behind modern standards. One classic example is the DPRK's introduction of its own Computer Numerical Control system in August 2010, to which Pyongyang boasted as a major technological breakthrough but was nevertheless on par with those produced by the US and the USSR in the 1960s. While there is a noticeable increase in the rate of automation and computerization, they are often combined with industrial machineries and practices that are still inferior in various technological aspects.

Issues are also created by the fact that many of the DPRK's key ordnance factories are built in the mountainous areas and/or underground. According to Kim Il-sung, the idea was to minimize disruptions and sustain the production and delivery of ordnances during wartime by protecting them from enemy bombardments.[41] While Kim Il-sung's rationale makes sense to a certain extent, the reality is that the rate of production in underground facilities is much lower than those on the ground, and the location of the factories in the mountainous areas also slows transportation. Thus, although the DPRK may have constructed industrial infrastructures that can work under wartime conditions, there are impediments in adequately producing and delivering goods to the KPA.

Despite some of the constraints, Kim Il-sung was serious about utilizing the heavy chemical industry to expand and modernize the military industry's technological capacity, calling for the "automation," "chemicalization," and "mechanization" of the KPA.[42] The DPRK's pride in military technologies is exhibited in places like the Museum of Arms and Equipment of the KPA as well as various forms of propaganda boasting how Pyongyang was able to establish its own military-industrial complex amid the disadvantageous circumstances. Yet the rhetoric is at odds with reality. Above all, the DPRK did not autonomously develop its military-industrial complex and technological capacity from bottom–up. The formation and development of the military industry drew much on the industrial assets left behind by the Japanese and more so the material and technical assistance from China and the USSR, as well as some Eastern European, Middle Eastern, and North African states. Additionally, the DPRK also benefitted from Korean returnees from China, Japan, and the USSR, and pro-DPRK Korean residents in Japan who provided industrial expertise and goods. More importantly, the myriad developments and issues in industrialization have resulted in mixed levels of autonomous R&D and production capacity of military technologies.

One area where the DPRK gained full autonomous capacity is small arms, mortars, and artillery systems. The DPRK was quick to become autonomous in small arms with the factories affiliated to the First General Bureau of the SEC manufacturing license production variants of Soviet automatic rifles such as the Type-58 based on the AK-47, Type-68 based on the AKM, Type-88 based on the AK-74, and also the Type-73 light machine gun influenced by the PK. Similar trends are also seen in handguns, mortars, grenades, and anti-personnel/tank mines which again are based on Soviet, East European, and Chinese models. Likewise, heavy weaponry such as artillery pieces and MLRS produced by the Third General Bureau has also become fully autonomous by the 1970s, evidenced by the large stock and variety of these weapons systems in the KPA inventory.

Another area where the DPRK's heavy industry capacity bore fruit is the automobile and vehicle industry. Since the late 1940s, the DPRK has imported a variety of vehicles from various states not only for immediate use but also to serve as technological templates for domestic production. Much of the vehicles came from the USSR/Russia and China, but a number of models were also imported from the selected East European states such as Czechoslovakia, East Germany, and Romania. Interestingly, a number of ground vehicles from non-communist states have also made their way to the DPRK, including cars, tractors, and trucks from Japan and West Germany. By the late 1950s, the DPRK was able to autonomously produce tractors and trucks with the establishment of factories such as the Tokchon Automobile Plant (now the Sungri Motor Complex) as well as the Kanggye Tractor Complex and Kiyang Tractor Complex. Similar developments were seen in railway platforms, where the DPRK not only inherited some platforms from Japan but also imported various types of electric and diesel locomotives from the USSR, China, Czechoslovakia, and others. By the 1960s, the Kim Jong Thae Electric Locomotive Complex was able to manufacture a variety of diesel and electric locomotives.

The DPRK's capacity to build heavy-duty vehicles significantly benefitted the indigenous construction of the KPA's armored and combat vehicles and also vehicles used for self-propelled and towed artillery and MLRS, as well as mobile missile launchers such as transporter-erector-launchers (TEL), mobile-erector-launchers (MEL), and transporter-erectors (TE). Indeed, the DPRK has continued to draw on foreign technologies, with a good number of the vehicles showcased at the various military parades and weapons tests closely resembling Chinese, Russian, Japanese, and European vehicles. Nevertheless, the DPRK has demonstrated its growing capacity in engineering various relatively modern variants of vehicles for the KPA.

In other areas, however, the DPRK's vehicle building capacity has been limited. In maritime platforms, the DPRK has a number of shipbuilding factories located near the port cities on the eastern and western coasts such as Chongjin, Nampho, Rajin, Sinpho, and Wonsan. Although the DPRK has constructed a number of large cargo and passenger vessels capable of sailing on open seas, the military vessels constructed by the factories of the Sixth General Bureau have been compact in size, with much of them being patrol or torpedo boats, and medium- and small-sized submarines. The technologies of much of the naval vessels constructed by the DPRK have been, and to a great extent still are based on the corvettes, frigates, patrol boats, and submarines imported from China and the USSR during the Cold War.

Greater problems are seen in aircraft where the factories under the Seventh General Bureau have been slow in mastering the ability to produce modern aerial platforms. Unlike China that gained the ability to build variants of Soviet aircraft and later its own models, the DPRK has struggled to go beyond the assembly production of vintage Soviet aircraft such as the An-2, Il-28, Mi-2, MiG-15, MiG-17, MiG-21, and Yak-18, and have faced difficulties with the more modern aircraft of the KPAAF inventory such as the MiG-23, MiG-29, and the Su-25.[43] Although the DPRK had originally signed an agreement with the USSR to license produce the MiG-29, this was later canceled after Russia stopped providing parts.[44] The DPRK's constraints in producing modern aircraft are much due to the fact that the technological complexity of aerial platforms far exceeds those of ground and naval vehicles and are also much less forgiving and flexible in producing variants. The only exception perhaps is in unmanned aerial systems (UAS), where the DPRK seems to have exploited technologies imported from China, Russia, and Syria to produce simple but workable drones.

Despite the mixed results in attaining self-sufficient production of vehicles, the DPRK has made significant strides in its capacity to produce ballistic missiles. The roots of the DPRK's missile program traces back to the 1960s when the USSR provided various tactical missiles including the S-75 (SA-2 Guideline), S-2

(SSC-2b Samlet), P-20 (SS-N-2 Styx), and the 3R10 (FROG-5).[45] The DPRK then moved onto ballistic missiles by acquiring SCUD missiles from Egypt in 1976 that enabled the indigenous production of short- and medium-range ballistic missiles. Then in the late 1990s, the DPRK demonstrated its quest to develop long-range ballistic missiles with the test launch of the Taepodong-1 that served as a technological demonstrator for future ballistic missiles. Significant developments have been evident since, with the construction of the Musudan BM-25 in the 2000s and then the advanced models of the Hwasong and Pukguksong series.

Another area where the DPRK has made much progress is chemical, biological, radiological, and nuclear weapons technologies. The DPRK's nuclear program started in the late 1950s when Moscow agreed to provide technical support for the establishment of a nuclear research facility.[46] Utilizing the technological know-how from the USSR, the DPRK constructed its first nuclear research reactor in North Pyongan Province in the 1960s, later named as the Nyongbyon Nuclear Scientific Research Center. Although the Pyongyang–Moscow nuclear agreement and Nyongbyong facility were ostensibly for energy-purposes, the developments nonetheless grew into the nuclear weapons program with the capability of producing plutonium and highly enriched uranium.

The DPRK's capacity to produce chemical and biological weapons is believed to have started in the mid-1950s when it established various science and technology institutions and programs focusing on biotechnology and chemistry.[47] While the DPRK has faced hurdles in other areas, its biotechnology and chemical programs are well established and staffed. Moreover, as industrialization advanced, the DPRK has built and imported machines from its benefactors (or smuggled from other states) that could be used to produce and preserve chemicals and pathogens. Yet much is unknown about the specific institutions and programs involved in the R&D of chemical and biological weapons as they could be clandestinely produced in the non-military sector such as in the various agricultural and medical facilities, as well as science laboratories of universities and other research institutions.

Mixed developments are also seen in the area of electronic and cyber warfare capabilities. The details of the DPRK's acquisition of radar and communication systems are unknown, although much would have come from the USSR and China during the Cold War. Kim Il-sung had often emphasized electronic warfare as a critical component of modernizing the KPA. The DPRK's push for electronic warfare capabilities was based on its strong sense of technological deficit, particularly after witnessing the equipment installed in the USS Pueblo seized in January 1968 and also various wars during and after the Cold War.[48] The actual progress in gaining autonomous capacity to produce advanced communications, radar, and sonar equipment had been slow, evidenced by the fact that the KPA still utilizes systems acquired from, or based on Soviet and Chinese models from the Cold War. Yet although the electronic warfare capabilities of the DPRK still lag behind advanced states, developments are evidenced by the operation of jamming systems as well as some developments in precision guidance and navigation systems.

The DPRK has also pursued advancements in space technologies for both civilian and military purposes. Sometime in the 1980s, the DPRK established the Korean Committee of Space Technology (currently known as the National Aerospace Development Administration) and in 1984 began its construction of a satellite communication center known as the Pyongyang Earth Station. Around the same time, the DPRK also embarked on its satellite program known as Kwangmyong-song. The first self-proclaimed launch took place in August 1998 when the DPRK launched the Taepodong-1, although there was no evidence of a satellite entering orbit. The DPRK also attempted to launch the Kwangmyongsong-2 and -3 in April 2009 and April 2012, although failed on both occasions. Then in December 2012, the DPRK successfully launched the Kwangmyongsong-3 into orbit, followed by the the Kwangmyongsong-4 in February 2016. Despite the eventual success in the DPRK's satellite launches, there are questions over the actual application and effectiveness of the satellites, with officials noting that the satellites have been "tumbling" in a low altitude orbit, making it incapable of functioning in an effective manner.[49]

Significant developments are seen in ICT. According to Bermudez, the DPRK first acquired mainframe computers in 1960s from the USSR and began producing its own computing systems in the 1980s.[50] Developments accelerated in the 1990s, with the establishment of various institutions for the training of ICT specialists, such as the Korea Computer Centre and the Pyongyang Program Centre, as well as a number of colleges and departments within universities that specialize in areas such as hardware and software engineering, communications, information science, machine learning, and others. By the 2000s, the DPRK began developing its own Linux-based operating system known as Pulgunbyol customized not only to state-sanctioned content but also to ensure greater levels of information security. Starting from the 2000s, the DPRK also made strides in hardware, with computers produced by Achim, Arirang-series smartphones, and Samjiyon and Ullim tablet computers. There were also developments in communication networks. In June 1997, the DPRK introduced a nation-wide intranet system known as Kwangmyong, linking the critical government infrastructures and institutions. The DPRK's communication sector continued to grow in the following years, with the installation of a 3G network in the late 2000s. While certainly not the most advanced, there is a notable increase in the level of computerization in the KPA, as seen in the military command centers, battlefield facilities and some portable tactical network systems. Still, there are questions over the level of advancement and penetration given the number of vintage systems that are incompatible with advanced digitalized technologies.

Although the numbered bureaus of the SEC appear to cover all types of systems, the DPRK is still yet to have the capacity to self-sufficiently produce all the equipment and platforms for the armed forces. Shin and Kim argue that by the 1990s the DPRK was able to self-sufficiently produce over 90% of its platforms with the exception of special mechanized units and precision instruments.[51] Yet the remaining 10% where the DPRK does not have autonomous capacity are technologies that are vital to complete the KPA's capabilities. Of the capabilities that the DPRK is capable of producing, much of them have been assembly/license productions, or crudely modified variants of Cold War Chinese and Soviet systems. Even regarding the new platforms showcased in recent years (such as in the 75th anniversary of the WPK in October 2020), almost all of them take after Chinese or Soviet/Russian models in some way. Moreover, even though there are developments in vehicles, there are questions over the equipment installed in those platforms to enable advanced combat functions.

Indeed, there is nothing wrong with importing and learning from foreign systems, and to the DPRK's credit they took the right steps in working with proven technologies. The problem, however, is that Pyongyang has suffered from inconsistent and incoherent access to technology from Beijing and Moscow during and after the Cold War. Technologically, both China and the USSR did not always provide their best systems to the DPRK with much of them being at least a generation old. Moreover, there were also geopolitical factors, where the Sino–Soviet split had a profound effect on the DPRK's procurement patterns. When Pyongyang leaned toward Beijing, it had to settle with technologies that were mostly Chinese variants of Soviet systems, or Chinese originals of compromised and nascent technological specifications.[52]

Problems in Moscow also affected the provision of military hardware to Pyongyang. The USSR did provide the DPRK with an array of technologies in the 1960s and then again in the 1980s with USD 2 billion worth of military aid including the MiG-23, MiG-29, and the Su-25.[53] But much changed after the fall of the USSR. Although Russia continued its efforts to produce various forms of advanced military hardware, it faced further economic troubles in the 1990s, making Moscow less generous and more demanding for any arms deals. In 2011, Kim Jong-il visited a Russian military aircraft factory in Ulan-Ude but proved to be more of a tour of the facility rather than any actual military transactions.[54] Thus, while the USSR was essential in the modernization of the KPA and the military industry, the level of technological transfers has been limited in the post-Cold War era.

To compensate for the loss of Moscow as the source of advanced military technologies, Pyongyang increasingly relied on Beijing. Yet hardware acquisitions from China have also proved to be particularly difficult in the post-Cold War era. The problem was not just due to China becoming increasingly transactional but also about its growing reluctance to provide arms to the DPRK that could trigger harsh reactions from the US, ROK, and Japan. One case of China's reluctance was seen in June 2010 when Hu Jintao reportedly declined Kim Jong-il's request to acquire the J-10 multi-role aircraft.[55] Still, the transfer of military technologies from China has continued to take place albeit on a smaller and more discreet scale. One example is the Wanshan WS51200 truck that is used by the DPRK not only as a TEL for ballistic missiles but also as a technological template to produce its own variants.[56]

Another major problem is that Pyongyang struggled to diversify its procurement network beyond Beijing and Moscow. The biggest reason is the limited number of states that were able to produce advanced weapons systems and also willing to trade with the DPRK. Other than China and the Soviet bloc, the only countries that provided the DPRK with substantial equipment during the Cold War were countries from the Middle East and North African regions. Consequently, the DPRK also turned to illegal imports via contractors or other routes. In the late 1990s, it was revealed that the DPRK had illegally imported a number of fighter jets including the MiG-21 and components of the MiG-29 from Kazakhstan. Some of the illicit deals also involved technologies from the West and Japan. For example, in the 1980s, the DPRK purchased the MD-500D/E helicopters via brokers in West Germany. In other cases, various Japanese electronic appliances, equipment, and parts have been shipped to the DPRK via pro-North Korean residents in Japan and other illicit brokers. While the DPRK was relatively successful in some of its underground procurements, their ability to continue the practice became increasingly difficult with the various sanctions and tightening of trade control measures. Moreover, illicit arms deals are often sophisticated and inconsistent, making them less reliable for states to strengthen their armed forces.

There are also questions regarding the DPRK's scientific and technological exchanges. During the Cold War, the USSR conducted numerous exchanges with the West that were then exploited for military applications. Likewise, the DPRK has also shown great interest in accessing technologies and know-how from both advanced and emerging technological powers. In particular, the DPRK held technological exchanges with the USSR from the early years and later China. Regarding nuclear power, the DPRK had sent scientists and technocrats to the USSR as early as the late 1950s, contributing to its long-term nuclear weapons project. Yet although North Korean academic and R&D institutions worked with foreign counterparts, the size and depth of the network, as well as their effects were nonetheless limited. The problems were not simply due to the increasingly stricter trade controls but also the reluctance of the counterparts in sharing their technologies for various political, economic, and technological reasons. To circumvent the constraints, the DPRK turned to clandestine deals and espionage to acquire technological know-how. One notable case was the clandestine deal with the Pakistani nuclear scientist Abdul Qadeer Khan who assisted the DPRK's nuclear program.

Of course, none of the above means that the DPRK is not getting anything at all. Modern technologies are trickling into the DPRK as evidenced by some of the more modern KPA platforms ranging from missiles to vehicles. Nevertheless, it is more accurate to say that the DPRK has merely picked up the pace in replacing the dated weapons with newer ones and is still in the early stages of the technological advancement curve. Even today, the large number of platforms in the KPA remains to be mix-and-match Frankenstein modifications of indigenously produced or imported technologies that are at least a generation or two old. Thus, although the creation of the military-industrial complex enhanced the DPRK's autonomous production capacity to some extent, the actual level of autonomous technological innovation still remain to be limited, particularly in the high-end platforms that utilize sophisticated automated technologies or command, control, communications, computers, intelligence, surveillance, target acquisition, and reconnaissance (C4ISTAR) systems.

From the cost-benefit viewpoint, there are many questions regarding the DPRK's quest to autonomously produce military assets. Eberstadt provides a sound assessment, arguing that the "North Korean defense industries presumably have embarked upon projects where their rates of return upon capital expenditures were extremely low; if so, it is possible that the resource requirements of the North Korean defense effort could have escalated suddenly and steeply" and despite the efforts to increase industrial and manpower capacity, "attempts at military modernization may be especially costly in an economy where technological innovation lags and international avenues of technology transfer are marginal."[57] While convincing, the limitations in technological capacity do not necessarily mean the DPRK completely failed, as some of the technological limitations are not simply about inability, but a matter of choice. Minus the bravado in publicized statements, the DPRK has been quite pragmatic and realistic about the state of their technological capacity, understanding that they are behind the technologically superior powers in indigenously producing and acquiring high-end systems. Nevertheless, Pyongyang has worked on building the industrial capacity for the capabilities that are of highest priority while restraining itself in areas that prove to be cost-inefficient. Prime examples would be long-range bombers and blue-water naval platforms, where the DPRK not only viewed these technologies as less necessary given their limited strategic scope, but more so because it would overstretch its industrial capacity, making it more sensible to make off-the-shelf purchases. Even though power projection was needed to deal with the US and Japan, the gap was filled by the pursuit of ballistic missile and WMD capabilities.

Human resource capacity

Over the years, the DPRK increased its dependence on and exploited its population to meet the military's human resource demands. Generally, citizens who reach the age of 14 are listed as potential draftees and would then undergo physical examinations at the age of 15. In principle, all candidates are expected to apply, though there are exceptions and exclusions. Exceptions (or deferments) are granted to individuals such as those in elite and/or specialized education institutes, selected artists and athletes, those with no siblings living with elderly parents, and those that can bribe themselves out. Exclusions apply to individuals who do not pass the enlistment tests, have convicted criminal records, as well as those from the lowest Songbun classes.

For the recruits, transitioning into military life is smoothened by the very nature of the North Korean society. Children live regimented lives in kindergarten and school, and those eligible are accepted into the Korean Children's Union and later the Kimilsungist–Kimjongilist Youth League. From the age of 14, students are mobilized into the RYG to serve as reservists and receive basic military skills (such as commands) to prepare them for their future tenure in the

KPA. Moreover, the aptitudes of the citizens are not as weak as one may assume, as the DPRK has a 12-year mandatory education system which includes military studies conducted during latter years of middle school that teaches the military application of biology, chemistry, first aid, geography, mathematics, physical education, and physics.[58]

Despite the comprehensive recruitment regime, there are serious problems in recruitment and retention caused by the deteriorating state of health, morale, and welfare among the populace. Unlike the past, the KPA today is losing its credibility as a vocation with prestige and value-add, particularly with the growing popularity of private entrepreneurship. Over the past few decades, many citizens have come to terms with the reality that a career in the military not only fails to provide daily essentials but also does not guarantee skillsets that would be useful in their post-military service careers. Even regarding the chances to join the WPK, many understand that such opportunities are on the condition that they make their way up the hierarchy which requires not only practical competence and political credibility but also in some cases bribes. Today, dodging military service is on the rise, either through legitimate or fabricated excuses or simply through bribes.

The widespread health problems, particularly those triggered in the 1990s and early 2000s, have forced Pyongyang to compromise its enlistment standards. Generally, the minimal standard to enlist has been 148 cm in height and 43 kg in weight. Yet the standard has been adjusted on numerous occasions as the DPRK began to recruit those who were born or grew up in the 1990s and 2000s. The most serious case was seen in 2010 when it was reported that the minimum physical requirements for enlistees had been lowered to 137 cm in height and 43 kg in weight.[59] The lowered standards indicate how the military was forced to scoop up a large number of individuals who may not be physically and mentally suitable for military duties.

Demographic problems have also significantly impacted recruitment. According to the UN Department of Economic and Social Affairs, the DPRK's birth rate has decreased significantly since the 1960s (less than two per family) while there has been a steady increase in the elderly.[60] Much was exacerbated by the widespread economic and human security crises during the "arduous march" era of 1990s, further denting the demographic patterns in the DPRK and leading to a drop in the number of individuals that can be recruited by the armed forces. According to Ishimaru, the ratio of new recruits to those discharging has gone down to 86% in recent years.[61] The problem is not simply about the number of personnel but also the imbalances in the structure of the KPA that consequently undermines its readiness.

Structural problems

Despite Kim Il-sung's all-in effort to establish the military–industrial complex and a war economy to autonomously strengthen the armed forces, his plans were not built on solid and sustainable industrial foundations. The military-industrial complex is not just about building factories and setting up a slush fund for the military sector but also about establishing a comprehensive and sustainable economic ecosystem. The North Korean model could be no more distant. At the Fifth Plenary Meeting of the Fourth CCWPK in December 1962, Kim Il-sung called for the parallel development of military readiness and the economy. Yet this essentially called for sacrifices on the civilian economy, where the 1963 Korea Central Yearbook stated, "[E]ven if the civilian (people's) economy is partially limited, military capability must first be strengthened."[62] Moreover, Kim Il-sung claimed that the civilian factories must be ready to produce goods to compensate shortfalls in the military factories.[63] Although the developments could be interpreted as the division of the economy into civilian and military spheres, it was also the creation of a zero-sum symbiotic economic system where the military sector garners a significant share of resources and technologies at the expense of the civilian sector.

The DPRK was not totally ignorant to the zero-sum problems caused. In his concluding speech at the Fifth WPK Congress on 12 November 1970, Kim Il-sung admitted that the extraordinarily high military outlays in the 1960s had in fact constrained state development – although he maintained that it was necessary given the circumstances that threatened the state's survival.[64] The high military outlays undoubtedly stressed the economy, particularly as the state's growth slipped into steady decline.[65] Still, Kim Il-sung continued to argue for greater developments in the military industry sector even as the economy plummeted into negative growth in the 1990s.[66] Kim Il-sung's logic was inherited by his successors, where Kim Jong-il declared that the state must continue its efforts in developing the military to survive the "arduous march."[67] Kim Jong-un has focused much more on economic development but nonetheless has paid particular attention on modernizing the KPA's capabilities by introducing the Byungjin line that was essentially a carbon copy of Kim Il-sung's "parallel development of economy and national defense capability" but with specific emphasis on nuclear weapons.

The DPRK continuously claimed that the development of the heavy industry sector is beneficial for the automation and mechanization of the light industry and agriculture sectors. Indeed, the heavy industry sector can help the light industry and agricultural sector, and the military industry too can contribute to the civilian sector through dual-use, spin-off, and leftover goods. Nevertheless, the benefits from the military-centric economy are only true under the right political, economic, and societal conditions. Even if one argues that the factories belonging to, or connected to the military sector may be better managed and therefore have better production capacity, reality shows that the actual benefits of the DPRK's military-centric economic policies in socio-economic development have been limited at best.[68] Kim Il-sung has in fact been aware of the gaps between military and civilian industrial products, where he complained at the National Conference on Quality Control Workers in February 1981 about how the quality of goods produced by the munitions sector is better than those in the civilian sector.[69] The zero-sum relationship between the military and civilian economies is not simply the result of military-centric policies but the lack of proper trade-off mechanisms to balance the two sectors.[70]

One also needs to consider the imbalance in the context of human resources. The general assumption would be that the mass armament of citizens would disrupt the labor force and consequently production. Even though one could argue that the military sector (both the KPA and military industry) provides employment, there are questions over the actual benefits to the civilian economy. In the last several decades, Pyongyang has mobilized military personnel for construction, factory, farming, and fishing labor. The system proved to be particularly vital during the 1990s and the 2000s, where Kim Jong-il praised the KPA for productively applying their war readiness to economic reconstruction.[71] The use of military personnel for labor is a double-edged sword. As Oh and Hassig note, the use of soldiers for labor had offset the zero-sum impact of the military-centric policies on the economy, although there are nonetheless questions concerning the level of expertise that contribute to genuine development.[72] Moreover, there are also problems for the military, where economic duties would inevitably undercut the time spent for training, consequently affecting the readiness of the KPA.

While the military sector has fared better than the others, it too has experienced decline. Ironically, as the armed forces and the military-industrial complex grew in size and became more prioritized, it also became increasingly vulnerable to the state's economic difficulties, where the capacity to finance and manage the KPA diminished in congruence with the state's resource base. Rather, the DPRK seems to have passed a certain threshold, where the concentrated development of the military sector has retarded the whole economy to the point where the civilian–military economic nexus has become negative-sum.[73] For instance, Sung et al. argued that during the 1990s, arms buildup in the army, navy, and air force contracted by 45%, 29%, and 87%, respectively.[74] Given that the 1990s onward was the period when the DPRK felt greater urgency to strengthen the KPA, Pyongyang faced the dilemma of how it can deal with the increasing demands despite the diminishing resources.

For the DPRK, economic growth is essential in strengthening its military industry. But the DPRK's economic growth much depends on governance and trade – which is made difficult by the self-imposed policies and self-inflicted problems. Despite the relative abundance of resources and industrialization in the early years that showed some potential for the DPRK to become a regional economic power, the obsession for regime survival and geopolitical agendas constrained it from converting itself into an export-oriented economy such as China, Japan, ROK, Singapore, and Taiwan. The failure to advance its trade profile undermined its ability to gain not only revenue but also resources that are domestically unavailable. The DPRK depended much on aid and concessions from China and the USSR as well as the selected East European states, but they were insufficient in sustainably nurturing the economy. Problems became greater in the post-Cold War era, leading to the scarcity of resources to revive and develop the troubled economy. At most, the DPRK attempted to attract foreign investment and trade by opening Special Economic Zones in Chongjin, Rajin–Sonbong, and Sinuiju, although they included rigid restrictions and poor level of transparency that deterred potential partners.

The myriad economic problems snowballed beyond the point of no return and the available remedies are few and far between. Although one could argue that boost in revenue would solve most of the DPRK's problems, much is dependent on two conditions. First, the DPRK would need to undertake some kind of economic reform. But while Kim Jong-un has indeed been forthcoming about economic development and has even encouraged market activities, the focus has been on modernizing the economy rather than reform. Second, boosting the DPRK's trade would require integration into the global economic community. However, the DPRK's trade faces severe constraints due to the economic sanctions against it, and the lifting of those measures would first require adequate steps by Pyongyang for denuclearization and also moderation of its bellicose behavior. Moreover, there are also quantitative and qualitative problems in the DPRK's transport and logistics infrastructures that create major bottlenecks in facilitating greater volumes of international trade.

There are also challenges in reforming the military industry to nurture the technological and production growth in a more competitive and free environment. Pyongyang's military industry is modeled on those of Beijing and Moscow in the 1950s and 1960s when they were centralized and closed. Yet since then, both Beijing and Moscow reformed their military-industrial complexes in accordance with the economic reforms that took place and also to focus more on technological advancements through better R&D networks (particularly in new and emerging technologies). Some kind of reform in the military industry sector would certainly benefit the DPRK in the production of more advanced systems. But there are issues regarding resources, and more so the regime's political reluctance to any forms of decentralization that could undermine its command and control of the military sector.

Impact on defense planning

Despite the establishment of the military-industrial complex and consistently allocating an extremely high amount of resources to the military sector, the actual returns have been limited at best. Much is due to not only the inherent resource and technology capacity issues but also the management system centered on the regime's centralized and politicized command and control than industrial rationales. Against this backdrop, high military outlays and the enormous size of the military-industrial complex have only exacerbated the inefficiencies and cost-ineffectiveness in the DPRK's defense planning that undermines not only the KPA's readiness but also the DPRK economy itself. Given the circumstances the DPRK faced, Kim Il-sung's establishment of an autonomous military-industrial complex was understandable. Pyongyang had never sought to become some kind of military technology powerhouse such as the US, USSR, or China but was simply seeking to attain a more consistent and reliable means of strengthening the KPA's readiness. Yet there were severe problems in the execution, where the

DPRK failed to establish the right economic and industrial ecosystem, as well as having the adequate access to resources and technologies.

The DPRK's economic and industrial capacity significantly impacted its defense planning. Above all, the DPRK has been forced to pursue the most affordable and feasible means within its boundaries to modernize its military capabilities in three ways. First, building upgraded capabilities based on both indigenous and imported models or designs. Second, the mix-and-match of weapons systems, instruments, and parts onto various ground, naval, and air vehicles to build and field a variety of capabilities. Third, extending the lifecycle of dated platforms to quantitatively enhance the KPA's readiness. Such measures have allowed the KPA to attain a diverse and quantitative stock of capabilities but nevertheless have created new problems such as constraints in the smooth succession of technologies, and also in creating an ecosystem and network of capabilities that are vital to enhance the readiness kill-chain. Furthermore, there are notable shortages in resources for O&M, particularly in petroleum, oil, lubricants, and electricity. Taken together, although there is no doubt that the establishment of the indigenous military-industrial complex and the heavy devotion of resources were critical for the KPA, many questions remain over the actual advancements in readiness. Of course, none of the above means that the DPRK cannot modernize or strengthen the KPA. As evidenced in recent years, genuine developments are taking place, meaning that the DPRK is investing more to modernize the armed forces. Moreover, Kim Jong-un has been serious about modernization, not just in military hardware and state infrastructures but also in revamping the education system with renewed emphasis on science and technology. Nevertheless, there are still many hurdles that inevitably constrain the level, rate, and volume of technological developments to meet the military readiness demands.

Notes

1 See: Il-sung Kim, "urineun jacheeui himeuro mugireul mandeuleo mujanghayeoya handa [We Must Make Weapons by Our Own Efforts to Arm Ourselves] (31 October 1949)," in *Kim Il Sung jeojakjib [Kim Il Sung Works]*, vol. 5 (Pyongyang, DPRK: Joseon Rodongdang Chulpansa, 1980), 297–99.

2 Masahiko Nakagawa, "chousenminsyusyugijinminkyouwakokuno gunjukougyou (1): kaihouchokugono gunmintenkanto gunjukougyouno kigen [DPRK's Military Industry (1): Civil-Military Conversion and the Foundations of the Military Industry in the Post-Liberation Period]," *Ajiken World Trend* 199 (April 2012): 56.

3 Ibid.

4 See: Il-sung Kim, "dangeul jiljeokeuro gongohi hamyeo gongeobsaengsane daehan dangjeokjidoreul gaeseonhalde daehayeo [On Ensuring Qualitative Consolidation of the Party and Improving Party Guidance of Industrial Production] (4 June 1953)," in *Kim Il Sung jeojakjib [Kim Il Sung Works]*, vol. 7 (Pyongyang, DPRK: Joseon Rodongdang Chulpansa, 1980), 498.

5 See: Zhihua Shen and Yafeng Xia, "China and the Post-War Reconstruction of North Korea, 1953–1961," in *North Korea International Documentation Project Working Paper Series* (Washington, DC: Woodrow Wilson International Center for Scholars, May, 2012).

6 See: Il-sung Kim, "byeonggigongeobeul deouk baljeonsikigi wihayeo [For Further Development of the Ordinance Industry] (28 May 1961)," in *Kim Il Sung jeojakjib [Kim Il Sung Works]*, vol. 15 (Pyongyang, DPRK: Joseon Rodongdang Chulpansa, 1981), 132, 137–46.

7 See: Il-sung Kim, "joguktongilwieobeul silhyeonhagi wihayeo hyeokmyeongryeokryangeul baekbangeuro ganghwahaja [Let Us Strengthen the Revolutionary Forces in Every Way so as to Achieve the Cause of Reunification of the Country] (27 February 1964)," in *Kim Il Sung jeojakjib [Kim Il Sung Works]*, vol. 18 (Pyongyang, DPRK: Joseon Rodongdang Chulpansa, 1982), 256–57.

8 See: Il-sung Kim, "joseonrodongdang je5cha daehoieseo han jungangwiwonhoisaeobchonghwabogo [Report to the Fifth Congress of the Workers' Party of Korea on the Work of the Central Committee] (2 November 1970)," in *Kim Il Sung jeojakjib [Kim Il Sung Works]*, vol. 25 (Pyongyang, DPRK: Joseon Rodongdang Chulpansa, 1983), 256–57.

9 See: Sung-bin Choi, Jae-moon Yoo, and Si-woo Kwak, *bukhan gunsusaneob gaehwang [Current State of the North Korean Military Industry]* (Seoul, ROK: Korea Institute for Defense Analyses, 2005), 28.

10 Numerous changes explained by Nakagawa Masahiko. See: Masahiko Nakagawa, "chousenminsyusyugijinminkyouwakokuno gunjukougyou (2): gunkeizaino seiritsu [DPRK's Military Industry (2): Establishment of the Military Economy]," *Ajiken World Trend* 201 (June 2012): 34–35. Kim Il-sung's statements on the First Bureau of Machine Industry can be found in his speech made on 5 August 1958. See: Il-sung Kim, "jagando dangdanchedeulape naseoneun myeotgaji gwaeobe daehayeo [Some Tasks of Party Organizations in Jagang Province] (5 August 1958)," in *Kim Il Sung jeojakjib [Kim Il Sung Works]*, vol. 12 (Pyongyang, DPRK: Joseon Rodongdang Chulpansa, 1981).

11 The name of the CCWPK Munitions Industry Department has been interchangeably renamed on numerous occasions. At the time of writing is known as the CCWPK Machine-Building Industry Department. Moreover, Nakagawa claims that the SEC was essentially based on the former First Bureau of the Ministry of Machine Building. See: Nakagawa, "chousenminsyusyugijinminkyouwakokuno gunjukougyou (2): gunkeizaino seiritsu [DPRK's Military Industry (2): Establishment of the Military Economy]," 35.

12 Korea Institute for National Unification, *bukhan gaeyo [Outline of North Korea]* (Seoul, ROK: Korea Institute for National Unification, 2009), 108; Kang-taeg Lim, "bukhaneui gunsusaneob jeongchaek [North Korea's Military Industry Policies]," in *bukhaneui gunsa [North Korean Military Affairs]*, ed. Bukhan Yeongu Hakhoi (Seoul, ROK: Gyeongin Munhwasa, 2006), 441–48; Chae-gi Sung, Joo-hyun Park, Jae-ok Park, and O-bong Kwon, *bukhan gyeongjewigi 10nyeongwa gunbijeunggang neungryeok [North Korea's Decade of Economic Crisis and Capacity for Military Buildup]* (Seoul, ROK: Korea Institute for Defense Analyses Press, 2003), 44–45.

13 Lim, "bukhaneui gunsusaneob jeongchaek [North Korea's Military Industry Policies]," 438; Joseph S. Bermudez Jr., *The Armed Forces of North Korea* (St. Leonards, Australia: Allen & Unwin, 2001), 45–55; US Senate Committee on Governmental Affairs Subcommittee on International Security, Proliferation and Federal Services, *Senate Hearing 105–241: North Korean Missile Proliferation,* October 21, 1997.

14 Lim, "bukhaneui gunsusaneob jeongchaek [North Korea's Military Industry Policies]," 448.

15 Yang-ju Kwon, *bukhangunsaeui ihae [The Comprehension of North Korean Military]*, Expanded ed. (Seoul, ROK: Korea Institute of Defense Analyses, 2014), 332.

16 Lim, "bukhaneui gunsusaneob jeongchaek [North Korea's Military Industry Policies]," 448.

17 Regarding governance on the state side, Lim argued that the SEC was governed by the NDC Bureau of Defense Planning. See: Ibid., 436.
18 Sung et al., *bukhan gyeongjewigi 10nyeongwa gunbijeunggang neungryeok [North Korea's Decade of Economic Crisis and Capacity for Military Buildup]*, 44.
19 Markus V. Garlauskas, "Ri Pyong Chol: Kim's New Right Hand Man?" *38 North*, August 5, 2020, www.38north.org/2020/08/mgarlauskas080520/.
20 Joseon Jungang Tongshinsa, *joseon jungang nyeongam 1966–1967 [Korea Central Yearbook 1966–1967]* (Pyongyang, DPRK: Joseon Jungang Tongshinsa, 1967), 114–15.
21 Taik-young Hamm, *Arming the Two Koreas: State, Capital and Military Power* (London, UK and New York: Routledge, 1999), 96.
22 Ibid., 58, 74.
23 US Department of State, "World Military Expenditures and Arms Transfers 2019," (2020), www.state.gov/world-military-expenditures-and-arms-transfers-2019/.
24 Hyo-joo Son, "N. Korean Military Spending Nearly 30% of S. Korea's: Defense Minister," *Dong-A Ilbo*, May 5, 2016, www.donga.com/en/article/all/20160505/533532/1/N-Korean-military-spending-nearly-30-of-S-Korea-s-defense-minister.
25 For instance, Moon and Lee argued that the DPRK's military expenditures increased from 3.3 billion KPW in 2002 to 50.7 billion KPW in 2003 largely due to Pyongyang's new fiscal and foreign exchanges policies in July 2002. Chung-in Moon and Sang-keun Lee, "Military Spending and the Arms Race on the Korean Peninsula," *Asian Perspective* 33, no. 4 (2009): 80.
26 Joseon Baekgwasajeon Pyeonchanwiwonhoi, *kwangmyongbaekgwasajeon [Kwangmyong Encyclopedia]* vol. 5 (Economics) (Pyongyang, DPRK: Baekgwasajeon Chulpansa, 2010): 264.
27 See: Chae-gi Sung, "bukhan gongpyogunsabi silchee daehan jeongmil jaebunseok [Detailed Re-examination of North Korea's Official Military Expenditures]," in *bukhaneui gunsa [North Korean Military Affairs]*, ed. Bukhan Yeongu Hakhoi (Seoul, ROK: Gyeongin Munhwasa, 2006), 465–69; Sung et al., *bukhan gyeongjewigi 10nyeongwa gunbijeunggang neungryeok [North Korea's Decade of Economic Crisis and Capacity for Military Buildup]*, 58–59, 62; Kwon, *bukhangunsaeui ihae [The Comprehension of North Korean Military]*, 285–88.
28 See: Sung et al., *bukhan gyeongjewigi 10nyeongwa gunbijeunggang neungryeok [North Korea's Decade of Economic Crisis and Capacity for Military Buildup]*, 61–63.
29 Lim, *bukhaneui gunsusaneob jeongchaeki gyeongjee michineun hyogwa bunseok [Analysis of the Economic Effects of the North Korean Military Industry]*, 67-68.
30 See: Sung et al., *bukhan gyeongjewigi 10nyeongwa gunbijeunggang neungryeok [North Korea's Decade of Economic Crisis and Capacity for Military Buildup]*, 44.
31 Masahiko Nakagawa, "chousenminsyusyugijinminkyouwakokuno gunjukougyou (3): gunkeizaito toukeizai [DPRK's Military Industry (3): The Military Economy and the Party's Economy]," *Ajiken World Trend* 204 (September 2012): 46–48; Sung et al., *bukhan gyeongjewigi 10nyeongwa gunbijeunggang neungryeok [North Korea's Decade of Economic Crisis and Capacity for Military Buildup]*, 41–43.
32 Ju-hwal Choi, "bukhanguneui woihwabeoli siltaewa jeonturyeoke michineun yeonghyang [Facts about the North Korean Military's Foreign Currency Earnings and their Impact on Military Capabilities]," *bukhanjosayeongu [Research on North Korea]* 2, no. 2 (1999): 27–28. Also see: Kwon, *bukhangunsaeui ihae [The Comprehension of North Korean Military]*, 296–99.
33 Nakagawa, "chousenminsyusyugijinminkyouwakokuno gunjukougyou (2): gunkeizaino seiritsu [DPRK's Military Industry (2): Establishment of the Military Economy]," 35–36.

34 Choi, "bukhanguneui woihwabeoli siltaewa jeonturyeoke michineun yeonghyang [Facts about the North Korean Military's Foreign Currency Earnings and their Impact on Military Capabilities]," 27–28.

35 See: Hideshi Takesada, "kitachousenno gunjiryokuwo saguru [Investigating North Korea's Military Capability]," in *kitachousen: sono jitsuzouto kiseki [North Korea: The Real Picture and its Path]* (Tokyo, Japan: Kobunken, 1998), 54; National Committee on North Korea, "An Overview of North Korea's Ballistic Missile Program," in *The National Committee on North Korea Issue Brief* (Washington, DC: National Committee on North Korea, 2011).

36 UN Security Council, "S/2010/571 Final Report of the Panel of Experts Submitted Pursuant to Resolution 1874," (UN Security Council, 5 November 2010), 27.

37 Kwon, *bukhangunsaeui ihae [The Comprehension of North Korean Military]*, 306.

38 For the list of companies, see: "S/2010/571 Final Report of the Panel of Experts Submitted Pursuant to Resolution 1874," (UN Security Council, 5 November 2010), 43. Also see the subsequent annual reports for detailed updates.

39 See: Bruce E. Jr. Bechtol, *North Korean Military Proliferation in the Middle East and Africa: Enabling Violence and Instability* (Lexington, KY: University Press of Kentucky, 2018).

40 Japan and ROK defense expenditures based on figures from the International Institute for Strategic Studies and Stockholm International Peace Research Institute. See: International Institute for Strategic Studies, *The Military Balance 2020* (London, UK: International Institute for Strategic Studies, 2020), 279, 286; Nan Tian, Alexandra Kuimova, Diego Lopes da Silva, Pieter D. Wezeman, and Siemon T. Wezeman, "Trends in Military Expenditure, 2019," in *Stockholm International Peace Research Institute Fact Sheet* (Stockholm, Sweden: Stockholm International Peace Research Institute, April 2020): 7–8.

41 See: Il-sung Kim, "inmingundaereul jiljeokeuro ganghwahayeo ganbugundaero mandeulja [Let Us Make the People's Army a Cadre Army by Strengthening it Qualitatively] (27 May 1954)," in *Kim Il Sung jeojakjib [Kim Il Sung Works]*, vol. 8 (Pyongyang, DPRK: Joseon Rodongdang Chulpansa, 1980), 433–34.

42 See: Il-sung Kim, "uri naraeui jeongsewa myeotgaji gunsagwaeobe daehayeo [On the Situation in Our Country and Some Military Tasks] (25 December 1961)," 627.

43 Bermudez Jr., *The Armed Forces of North Korea*, 57; Choi et al., "bukhan gunsusaneob gaehwang [Current State of the North Korean Military Industry]," 31; Beom-chul Shin and Jin-a Kim, "bukhanguneui unyeonggwa gunsusaneob [North Korea's Military Management and the Military Industry]," in *bukhangun sikeurit ripoteu [North Korea Military Secret Report]*, ed. Yong-won Yoo, Beom-chul Shin, and Jin-a Kim (Seoul, ROK: Planet Media, 2013), 369.

44 Choi et al., "bukhan gunsusaneob gaehwang [Current State of the North Korean Military Industry]," 32.

45 Bermudez Jr., *The Armed Forces of North Korea*, 237–39.

46 Balazs Szalontai and Sergey Radchenko, "North Korea's Efforts to Acquire Nuclear Technology and Nuclear Weapons: Evidence from Russian and Hungarian Archives," in *Cold War International History Project* (Washington, DC: Woodrow Wilson International Center for Scholars, 2006).

47 For the history of the DPRK's chemical and biological weapons programs, see: International Crisis Group, "North Korea's Chemical and Biological Weapons Programs," in *Asia Report* (Seoul, ROK and Brussels, Belgium: International Crisis Group, 2009).

48 Joseph S. Bermudez Jr., "SIGINT, EW, and EIW in the Korean People's Army: An Overview of Development and Organization," in *Bytes and Bullets in Korea*, ed. Alexandre Y. Mansourov (Honolulu, HI: APCSS, 2005), 236–43.

49 Andrea Shalal and Idrees Ali, "North Korea Satellite Tumbling in Orbit Again: U.S. Sources," *Reuters*, February 19, 2016, www.reuters.com/article/us-northkorea-satellite-idUSKCN0VR2R3.
50 Bermudez Jr., "SIGINT, EW, and EIW in the Korean People's Army: An Overview of Development and Organization," 238, 246.
51 Shin and Kim, "bukhanguneui unyeonggwa gunsusaneob [North Korea's Military Management and the Military Industry]," 363.
52 The DPRK's efforts to fill the gaps created were explained by Oberdorfer: "According to the account of a former Chinese official, a very senior North Korean military official, probably Defense Minister O Jin U, sought to match the Soviet weaponry with Chinese weaponry in the mid-1980s, making extensive requests for ships, planes, and other major weapons during an unpublicized trip to Beijing." Don Oberdorfer, *The Two Koreas: A Contemporary History* (New York: Basic Books, 2001), 504.
53 Hamm, *Arming the Two Koreas: State, Capital and Military Power*, 85.
54 Yong-won Yoo and Yong-su Lee, "junge toejja matdeoni . . . buk, ibeonen reosiae jeontugi dallago haetna [After being Rejected by China, Has North Korea Now Asked Russia for Fighter Planes?]," *Chosun Ilbo*, August 29, 2011, http://news.chosun.com/site/data/html_dir/2011/08/29/2011082900092.html.
55 Cheol-hwan Kang and Yong-hyun An, "Kim Jong-il: jungguke choisinye jeontugi jiwon yocheong [Kim Jong-il: Requests Provision of Modern Fighter Jets from China]," ibid., June 17, 2010, http://nk.chosun.com/news/news.html?ACT=detail&res_id=126058.
56 See: Jeffrey Lewis, Melissa Hanham, and Amber Lee, "That Ain't My Truck: Where North Korea Assembled Its Chinese Transporter-Erector-Launchers," *38 North*, February 3, 2014, 38north.org/2014/02/jlewis020314/.
57 Nicholas Eberstadt, "Development, Structure and Performance of the DPRK Economy: Empirical Hints," in *North Korea in Transition: Prospects for Economic and Social Reform*, ed. Lawrence J. Lau and Chang-ho Yoon (London, UK and Northampton, MA: Edward Elgar, 2001), 42.
58 For examples of middle school military studies textbooks, see: Yun-do Han, *gunsa jisik: junghakkyo je5,6haknyeonyong [Military Knowledge: For Fifth and Sixth Year Middle School Students]* (Pyongyang, DPRK: Gyoyukdoseo Chulpansa, 2006).
59 Yeong-tae Jeong, "bukhangun ibdaegijun 137cm. namcheukgwa 'meori hana' chai [North Korean Military Enlistment Standard at 137cm . . . Difference by 'One Head' with the South]," *SBS News*, March 19, 2010, https://news.sbs.co.kr/news/endPage.do?news_id=N1000723246.
60 See: UN Department of Economic and Social Affairs, *World Population Prospects: The 2017 Revision*, vol. 2 (Demographic Profiles) (New York: United Nations, 2017). North Korea was relatively candid in the 2008 Population Census about its demographic problems. Central Bureau of Statistics, "DPR Korea 2008 Population Census National Report," (Pyongyang, DPRK2009).
61 Jiro Ishimaru, "<N.Korea> N.Korean Military Facing Recruit Shortage Due to Increased Draft Dodgers and Deserters," *Rimjing-Gang*, January 24, 2017, www.asiapress.org/rimjin-gang/2017/01/military/20170120-military-service-evasion/.
62 Joseon Jungang Tongshinsa, *joseon jungang nyeongam 1963 [Korea Central Yearbook 1963]* (Pyongyang, DPRK: Joseon Jungang Tongshinsa, 1963), 159.
63 Kim, "byeonggigongeobeul deouk baljeonsikigi wihayeo [For Further Development of the Ordinance Industry] (28 May 1961)," 120.
64 Il-sung Kim, "joseonrodongdang je5cha daehoieseo han gyeolron [Concluding Speech at the Fifth Congress of the Workers' Party of Korea] (12 November 1970)," 219.

65 For detailed discussions on the DPRK's defense burden and economic growth, see: Hamm, *Arming the Two Koreas: State, Capital and Military Power*, 137–46.
66 See: Il-sung Kim, "dangmyeonhan sahoijueuigyeongjegeonseolbanghyange daehayeo [On the Direction of Socialist Economic Construction for the Immediate Period Ahead] (8 December 1993)," in *Kim Il Sung jeojakjib [Kim Il Sung Works]*, vol. 44 (Pyongyang, DPRK: Joseon Rodongdang Chulpansa, 1996), 284.
67 Kim Jong-il stated, "It is imperative to expedite overall economic construction while giving precedence to the defence industry . . . so as to support the Party's Songun politics both materially and technologically, and radically improve the people's living standards in a short period of time." See: Jong-il Kim, "songunhyeokmyeongroseoneun uri sidaeeui widaehan hyeokmyeongroseonimyeo uri hyeokmyeongeui baekjeonbaekseungeui gichiida [The Songun-Based Revolutionary Line is a Great Revolutionary Line of Our Era and an Ever-Victorious Banner of Our Revolution] (29 January 2003)," in *Kim Jong Il seonjib [Kim Jong Il Selected Works]*, vol. 21, Expanded ed. (Pyongyang, DPRK: Joseon Rodongdang Chulpansa, 2013), 393.
68 Kang-taeg Lim, *bukhaneui gunsusaneob jeongchaeki gyeongjee michineun hyogwa bunseok [Analysis of the Economic Effects of the North Korean Military Industry]* (Seoul, ROK: Korea Institute for National Unification, 2001), 84–86.
69 See: Il-sung Kim, "pumjilgamdoksaeobeul gaeseonganghwahalde daehayeo [On Improving Quality Control] (2 February 1981)," in *Kim Il Sung jeojakjib [Kim Il Sung Works]*, vol. 36 (Pyongyang, DPRK: Joseon Rodongdang Chulpansa, 1990).
70 Lim, *bukhaneui gunsusaneob jeongchaeki gyeongjee michineun hyogwa bunseok [Analysis of the Economic Effects of the North Korean Military Industry]*, 23.
71 See: Jong-il Kim, "inmingunjihwiseongwondeuleun nopeun yeoljeongeul jinigo hyeokmyeongjeokeuro, jeontujeokeuro ilhayeoya handa [Commanders of the People's Army Must Uphold Strong Passion and Work in Revolutionary and Militant Ways] (9 Oct 2004)," in *Kim Jong Il seonjib [Kim Jong Il Selected Works]*, vol. 22, Expanded ed. (Pyongyang, DPRK: Joseon Rodongdang Chulpansa, 2013), 146–49.
72 Kong-dan Oh and Ralph C. Hassig, *North Korea Through the Looking Glass* (Washington, DC: Brookings Institution Press, 2000), 115.
73 Hyeon-su Jung, Young-hwan Kim, and Wae-sul Kim, *bukhan jeongchigyeongjeron [Political-Economic Concepts of North Korea]* (Seoul, ROK: Sinyeongsa, 1995), 238.
74 Sung et al., *bukhan gyeongjewigi 10nyeongwa gunbijeunggang neungryeok [North Korea's Decade of Economic Crisis and Capacity for Military Buildup]*, 35–37, 46.

References

Bechtol Jr., Bruce E. *North Korean Military Proliferation in the Middle East and Africa: Enabling Violence and Instability*. Lexington, KY: University Press of Kentucky, 2018.

Bermudez Jr., Joseph S. *The Armed Forces of North Korea*. St. Leonards, Australia: Allen & Unwin, 2001.

———. "SIGINT, EW, and EIW in the Korean People's Army: An Overview of Development and Organization." In *Bytes and Bullets in Korea*, edited by Alexandre Y. Mansourov. Honolulu, HI: APCSS, 2005.

Central Bureau of Statistics. *DPR Korea 2008 Population Census National Report*. Pyongyang, DPRK: Central Bureau of Statistics, 2009.

Choi, Ju-hwal. "bukhanguneui woihwabeoli siltaewa jeonturyeoke michineun yeonghyang [Facts about the North Korean Military's Foreign Currency Earnings and Their Impact on Military Capabilities]." *bukhanjosayeongu [Research on North Korea]* 2, no. 2 (1999).

Choi, Sung-bin, Jae-moon Yoo, and Si-woo Kwak. *bukhan gunsusaneob gaehwang [Current State of the North Korean Military Industry]*. Seoul, ROK: Korea Institute for Defense Analyses, 2005.

Eberstadt, Nicholas. "Development, Structure and Performance of the DPRK Economy: Empirical Hints." In *North Korea in Transition: Prospects for Economic and Social Reform*, edited by Lawrence J. Lau and Chang-ho Yoon, xvii, 339p. London, UK and Northampton, MA: Edward Elgar, 2001.

Garlauskas, Markus V. "Ri Pyong Chol: Kim's New Right Hand Man?" *38 North*, August 5, 2020. www.38north.org/2020/08/mgarlauskas080520/.

Hamm, Taik-young. *Arming the Two Koreas: State, Capital and Military Power*. London, UK and New York, NY: Routledge, 1999.

Han, Yun-do. *gunsa jisik: junghakkyo je5,6haknyeonyong [Military Knowledge: For Fifth and Sixth Year Middle School Students]*. Pyongyang, DPRK: Gyoyukdoseo Chulpansa, 2006.

International Crisis Group. "North Korea's Chemical and Biological Weapons Programs." In *Asia Report*. Seoul, ROK and Brussels, Belgium: International Crisis Group, 2009.

International Institute for Strategic Studies. *The Military Balance 2020*. London, UK: International Institute for Strategic Studies, 2020.

Ishimaru, Jiro. "<N.Korea> N.Korean Military Facing Recruit Shortage Due to Increased Draft Dodgers and Deserters." *Rimjing-Gang*, January 24, 2017. www.asiapress.org/rimjin-gang/2017/01/military/20170120-military-service-evasion/.

Jeong, Yeong-tae. "bukhangun ibdaegijun 137cm. namcheukgwa 'meori hana' chai [North Korean Military Enlistment Standard at 137cm . . . Difference by 'One Head' with the South]." *SBS News*, March 19, 2010. https://news.sbs.co.kr/news/endPage.do?news_id=N1000723246.

Joseon Baekgwasajeon Pyeonchanwiwonhoi. *kwangmyongbaekgwasajeon [Kwangmyong Encyclopedia]*. Vol. 5 (Economics). Pyongyang, DPRK: Baekgwasajeon Chulpansa, 2010.

Joseon Jungang Tongshinsa. *joseon jungang nyeongam 1963 [Korea Central Yearbook 1963]*. Pyongyang, DPRK: Joseon Jungang Tongshinsa, 1963.

———. *joseon jungang nyeongam 1966–1967 [Korea Central Yearbook 1966–1967]*. Pyongyang, DPRK: Joseon Jungang Tongshinsa, 1967.

Jung, Hyeon-su, Young-hwan Kim, and Wae-sul Kim. *bukhan jeongchigyeongjeron [Political-Economic Concepts of North Korea]*. Seoul, ROK: Sinyeongsa, 1995.

Kang, Cheol-hwan, and Yong-hyun An. "Kim Jong-il: jungguke choisinye jeontugi jiwon yocheong [Kim Jong-il: Requests Provision of Modern Fighter Jets from China]." *Chosun Ilbo*, June 17, 2010. http://nk.chosun.com/news/news.html?ACT=detail&res_id=126058.

Kim, Il-sung. "dangeul jiljeokeuro gongohi hamyeo gongeobsaengsane daehan dangjeokjidoreul gaeseonhalde daehayeo [On Ensuring Qualitative Consolidation of the Party and Improving Party Guidance of Industrial Production] (4 June 1953)." In *Kim Il Sung jeojakjib [Kim Il Sung Works]*. Vol. 7. Pyongyang, DPRK: Joseon Rodongdang Chulpansa, 1980a.

———. "inmingundaereul jiljeokeuro ganghwahayeo ganbugundaero mandeulja [Let Us Make the People's Army a Cadre Army by Strengthening it Qualitatively] (27 May 1954)." In *Kim Il Sung jeojakjib [Kim Il Sung Works]*. Vol. 8. Pyongyang, DPRK: Joseon Rodongdang Chulpansa, 1980b.
———. "urineun jacheeui himeuro mugireul mandeuleo mujanghayeoya handa [We Must Make Weapons by Our Own Efforts to Arm Ourselves] (31 October 1949)." In *Kim Il Sung jeojakjib [Kim Il Sung Works]*. Vol. 5. Pyongyang, DPRK: Joseon Rodongdang Chulpansa, 1980c.
———. "byeonggigongeobeul deouk baljeonsikigi wihayeo [For Further Development of the Ordinance Industry] (28 May 1961)." In *Kim Il Sung jeojakjib [Kim Il Sung Works]*. Vol. 15. Pyongyang, DPRK: Joseon Rodongdang Chulpansa, 1981a.
———. "jagando dangdanchedeulape naseoneun myeotgaji gwaeobe daehayeo [Some Tasks of Party Organizations in Jagang Province] (5 August 1958)." In *Kim Il Sung jeojakjib [Kim Il Sung Works]*. Vol. 12. Pyongyang, DPRK: Joseon Rodongdang Chulpansa, 1981b.
———. "uri naraeui jeongsewa myeotgaji gunsagwaeobe daehayeo [On the Situation in Our Country and Some Military Tasks] (25 December 1961)." In *Kim Il Sung jeojakjib [Kim Il Sung Works]*. Vol. 15. Pyongyang, DPRK: Joseon Rodongdang Chulpansa, 1981c.
———. "joguktongilwieobeul silhyeonhagi wihayeo hyeokmyeongryeokryangeul baekbangeuro ganghwahaja [Let Us Strengthen the Revolutionary Forces in Every Way so as to Achieve the Cause of Reunification of the Country] (27 February 1964)." In *Kim Il Sung jeojakjib [Kim Il Sung Works]*. Vol. 18. Pyongyang, DPRK: Joseon Rodongdang Chulpansa, 1982.
———. "joseonrodongdang je5cha daehoieseo han gyeolron [Concluding Speech at the Fifth Congress of the Workers' Party of Korea] (12 November 1970)." In *Kim Il Sung jeojakjib [Kim Il Sung Works]*. Vol. 25. Pyongyang, DPRK: Joseon Rodongdang Chulpansa, 1983a.
———. "joseonrodongdang je5cha daehoieseo han jungangwiwonhoisaeobchonghwabogo [Report to the Fifth Congress of the Workers' Party of Korea on the Work of the Central Committee] (2 November 1970)." In *Kim Il Sung jeojakjib [Kim Il Sung Works]*. Vol. 25. Pyongyang, DPRK: Joseon Rodongdang Chulpansa, 1983b.
———. "pumjilgamdoksaeobeul gaeseonganghwahalde daehayeo [On Improving Quality Control] (2 February 1981)." In *Kim Il Sung jeojakjib [Kim Il Sung Works]*. Vol. 36. Pyongyang, DPRK: Joseon Rodongdang Chulpansa, 1990.
———. "dangmyeonhan sahoijueuigyeongjegeonseolbanghyange daehayeo [On the Direction of Socialist Economic Construction for the Immediate Period Ahead] (8 December 1993)." In *Kim Il Sung jeojakjib [Kim Il Sung Works]*. Vol. 44. Pyongyang, DPRK: Joseon Rodongdang Chulpansa, 1996.
Kim, Jong-il. "inmingunjihwiseongwondeuleun nopeun yeoljeongeul jinigo hyeokmyeongjeokeuro, jeontujeokeuro ilhayeoya handa [Commanders of the People's Army Must Uphold Strong Passion and Work in Revolutionary and Militant Ways] (9 Oct 2004)." In *Kim Jong Il seonjib [Kim Jong Il Selected Works]*. Vol. 22. Expanded ed. Pyongyang, DPRK: Joseon Rodongdang Chulpansa, 2013a.
———. "songunhyeokmyeong roseoneun uri sidaeeui widaehan hyeokmyeongroseonimyeo uri hyeokmyeongeui baekjeonbaekseungeui gichiida [The Songun-Based Revolutionary Line is a Great Revolutionary Line of Our Era and an Ever-Victorious Banner of Our Revolution] (29 January 2003)." In *Kim Jong Il seonjib*

[Kim Jong Il Selected Works]. Vol. 21. Expanded ed. Pyongyang, DPRK: Joseon Rodongdang Chulpansa, 2013b.

Korea Institute for National Unification. *bukhan gaeyo [Outline of North Korea]*. Seoul, ROK: Korea Institute for National Unification, 2009.

Kwon, Yang-ju. *bukhangunsaeui ihae [The Comprehension of North Korean Military]*. Expanded ed. Seoul, ROK: Korea Institute of Defense Analyses, 2014.

Lewis, Jeffrey, Melissa Hanham, and Amber Lee. "That Ain't My Truck: Where North Korea Assembled Its Chinese Transporter-Erector-Launchers." *38 North*, February 3, 2014. 38north.org/2014/02/jlewis020314/.

Lim, Kang-taeg. *bukhaneui gunsusaneob jeongchaeki gyeongjee michineun hyogwa bunseok [Analysis of the Economic Effects of the North Korean Military Industry]*. Seoul, ROK: Korea Institute for National Unification, 2001.

———. "bukhaneui gunsusaneob jeongchaek [North Korea's Military Industry Policies]." In *bukhaneui gunsa [North Korean Military Affairs]*, edited by Bukhan Yeongu Hakhoi. Seoul, ROK: Gyeongin Munhwasa, 2006.

Moon, Chung-in, and Sang-keun Lee. "Military Spending and the Arms Race on the Korean Peninsula." *Asian Perspective* 33, no. 4 (2009): III–69.

Nakagawa, Masahiko. "chousenminsyusyugijinminkyouwakokuno gunjukougyou (1): kaihouchokugono gunmintenkanto gunjukougyouno kigen [DPRK's Military Industry (1): Civil-Military Conversion and the Foundations of the Military Industry in the Post-Liberation Period]." *Ajiken World Trend* 199 (April 2012a).

———. "chousenminsyusyugijinminkyouwakokuno gunjukougyou (2): gunkeizaino seiritsu [DPRK's Military Industry (2): Establishment of the Military Economy]." *Ajiken World Trend* 201 (June 2012b).

———. "chousenminsyusyugijinminkyouwakokuno gunjukougyou (3): gunkeizaito toukeizai [DPRK's Military Industry (3): The Military Economy and the Party's Economy]." *Ajiken World Trend* 204 (September 2012c).

National Committee on North Korea. *"An Overview of North Korea's Ballistic Missile Program."* In *The National Committee on North Korea Issue Brief*, Washington, DC: National Committee on North Korea, 2011. https://www.ncnk.org/sites/default/files/NCNK_Issue_Brief_Ballistic_Missiles.pdf

Oberdorfer, Don. *The Two Koreas: A Contemporary History*. New York, NY: Basic Books, 2001.

Oh, Kong-dan, and Ralph C. Hassig. *North Korea Through the Looking Glass*. Washington, DC: Brookings Institution Press, 2000.

Shalal, Andrea, and Idrees Ali. "North Korea Satellite Tumbling in Orbit Again: U.S. Sources." *Reuters*, February 19, 2016. www.reuters.com/article/us-northkorea-satellite-idUSKCN0VR2R3.

Shen, Zhihua, and Yafeng Xia. "China and the Post-War Reconstruction of North Korea, 1953–1961." In *North Korea International Documentation Project Working Paper Series*. Washington, DC: Woodrow Wilson International Center for Scholars, May 2012.

Shin, Beom-chul, and Jin-a Kim. "bukhanguneui unyeonggwa gunsusaneob [North Korea's Military Management and the Military Industry]." In *bukhangun sikeurit ripoteu [North Korea Military Secret Report]*, edited by Yong-won Yoo, Beom-chul Shin, and Jin-a Kim. Seoul, ROK: Planet Media, 2013.

Son, Hyo-joo. "N. Korean Military Spending Nearly 30% of S. Korea's: Defense Minister." *Dong-A Ilbo*, May 5, 2016. www.donga.com/en/article/all/20160505/533532/1/N-Korean-military-spending-nearly-30-of-S-Korea-s-defense-minister.

Sung, Chae-gi. "bukhan gongpyogunsabi silchee daehan jeongmil jaebunseok [Detailed Re-examination of North Korea's Official Military Expenditures]." In *bukhaneui gunsa [North Korean Military Affairs]*, edited by Bukhan Yeongu Hakhoi. Seoul, ROK: Gyeongin Munhwasa, 2006.

Sung, Chae-gi, Joo-hyun Park, Jae-ok Park, and O-bong Kwon. *bukhan gyeongjewigi 10nyeongwa gunbijeunggang neungryeok [North Korea's Decade of Economic Crisis and Capacity for Military Buildup]*. Seoul, ROK: Korea Institute for Defense Analyses Press, 2003.

Szalontai, Balazs, and Sergey Radchenko. "North Korea's Efforts to Acquire Nuclear Technology and Nuclear Weapons: Evidence from Russian and Hungarian Archives." In *Cold War International History Project*. Washington, DC: Woodrow Wilson International Center for Scholars, 2006.

Takesada, Hideshi. "kitachousenno gunjiryokuwo saguru [Investigating North Korea's Military Capability]." In *kitachousen: sono jitsuzouto kiseki [North Korea: The Real Picture and its Path]*. Tokyo, Japan: Kobunken, 1998.

Tian, Nan, Alexandra Kuimova, Diego Lopes da Silva, Pieter D. Wezeman, and Siemon T. Wezeman. "Trends in Military Expenditure, 2019." In *Stockholm International Peace Research Institute Fact Sheet*. Stockholm, Sweden: Stockholm International Peace Research Institute, April 2020.

UN Department of Economic and Social Affairs. *World Population Prospects: The 2017 Revision*. Vol. 2 (Demographic Profiles). New York, NY: United Nations, 2017.

UN Security Council. "S/2010/571 Final Report of the Panel of Experts Submitted Pursuant to Resolution 1874" UN Security Council, November 5, 2010.

US Department of State. "World Military Expenditures and Arms Transfers 2019." (2020). www.state.gov/world-military-expenditures-and-arms-transfers-2019/.

US Senate Committee on Governmental Affairs Subcommittee on International Security, Proliferation and Federal Services. *Senate Hearing 105–241: North Korean Missile Proliferation*, October 21, 1997.

Yoo, Yong-won, and Yong-su Lee. "junge toejja matdeoni . . . buk, ibeonen reosiae jeontugi dallago haetna [After Being Rejected by China, Has North Korea Now Asked Russia for Fighter Planes?]." *Chosun Ilbo*, August 29, 2011. http://news.chosun.com/site/data/html_dir/2011/08/29/2011082900092.html.

5 The defense planning framework

For any state, defense planning is fraught with dilemmas, especially when there is a zero-sum balance between the strategic interests, political characteristics, and economic capacity. Such problems not only impact how the state's military is armed, but also how they are managed and operated. The DPRK presents an interesting case, where it aimed to kill two birds with one stone – autonomously strengthen the readiness of the armed forces while also bolstering the regime's absolute command and control despite the limited economic and industrial capacity. The defense planning framework had always loomed since the formative years but was embodied in the 1960s when Kim Il-sung introduced the Line of Self-Reliant Defence that aimed to: establish a cadre army; modernize the whole armed forces; arm the whole populace; and fortify the whole country. Although the doctrine instigated new developments in the KPA, it was not some kind military reform program. Rather, the doctrine was the embodiment of Kim Il-sung's concept to build a self-reliant armed forces while enhancing the regime's centralized and politicized command and control. However, Kim Il-sung's comprehensive and holistic approach to kill two birds with one stone also forced the DPRK's defense planning to work in a rigid framework replete with dilemmas, where the lack of resources and access to technology combined with the regime's obsession with its survival inevitably affected the KPA's readiness.

The DPRK's defense planning doctrine: Line of Self-Reliant Defence

According to the DPRK, Kim Il-sung was determined to establish a self-reliant military from his time as an anti-Japanese partisan as the key means of guaranteeing Korea's independence and sovereignty.[1] Almost immediately after the liberation of the Korean peninsula, Kim Il-sung led his colleagues to establish quasi-military institutions that would later become the KPA in 1948. Up to the Korean War, the DPRK had numerous advantages over the ROK in the military balance, having built a sizable, well-armed force. The KPA gained significant material and technical assistance from the USSR and were also relatively well manned. In contrast, the ROK had barely put together a combat-ready military, and the US forces had already withdrawn from the Korean peninsula by

June 1949. Yet despite the advantages, Kim Il-sung's goal of unifying the Korean peninsula was denied as the UN forces rolled the KPA back up north in the fall of 1950. At the Third Plenary Meeting of the CCWPK in December 1950, Kim Il-sung extensively outlined the reasons for the troubled military campaign and called for the enhancement of the cadres' leadership skills, further production and better utilization of weaponry, and armament of citizens.[2]

By the time the armistice was signed on 27 July 1953, the KPA was heavily bruised and reconstructing it was going to be extremely expensive. Yet there were a number of internal and external developments that were concerning for Kim Il-sung, upping the need to revamp the armed forces. Internally, Kim Il-sung was fighting to maintain and strengthen his rule, including the massive purges in the 1950s. Externally, there were concerns about the US, ROK, and Japan's security postures, and also questionable developments within and between China and the USSR. The changing strategic environment created greater vulnerabilities for the DPRK, leading to reconfigurations in its defense planning under the auspices of Juche. Kim Il-sung's efforts to implement the new defense planning doctrine came in the early 1960s with the introduction of the Line of Self-Reliant Defence. At the Fifth Plenary Meeting of the Fourth CCWPK in December 1962, Kim Il-sung called for the: establishment of a cadre army; arming of the populace; and fortification of the whole country. Then at the Second WPK Conference in October 1966, the DPRK added another guideline – "modernization of the whole forces." The late addition of "military modernization" was a renewed emphasis on improving the KPA's assets for modern warfare by establishing an indigenous military–industrial complex. The contents of the Line of Self-Reliant Defence, however, was nothing new, as Kim Il-sung had already talked about the individual guidelines in various separate speeches during the 1950s. Thus, the introduction of the Line of Self-Reliant Defence was essentially about packaging the four guidelines as a single defense planning doctrine to autonomously strengthen the KPA's readiness while also bolstering the leadership's command and control over the armed forces.

A key characteristic of the Line of Self-Reliant Defence is how it grew beyond a one-off policy and was later constitutionalized as the pillar of national defense. Article 60 of the current Socialist Constitution reads, "The State shall implement the Line of Self-Reliant Defence, the import of which is to train the army to be a cadre army, modernize the army, arm all the people and fortify the country on the basis of equipping the army and the people politically and ideologically." However, the codification of the Line of Self-Reliant Defence into the Socialist Constitution was incremental. The revised constitution in 1972 barely went further than stating that the government "shall implement the Line of Self-Reliant Defence," and the contents of the doctrine were not mentioned until the next round of constitutional revision in 1992. The three-decade gap between the introduction of the Line of Self-Reliant Defence in the 1960s and the full constitutional codification in 1992 suggests that the DPRK wanted to achieve the doctrinal goals to some degree before featuring it in the constitution. Such practices are common in the DPRK, just like how the Juche idea was first mentioned in the mid-1950s but not given full official status as the sole, guiding ideology until a more than a decade later. Nevertheless, the constitutional codification of the Line of Self-Reliant Defence indicates not only the DPRK's faith in the doctrine as the permanent defense planning directive but also the high level of path-dependence with limited room for revision.

Despite the fixed nature of the Line of Self-Reliant Defence, the specifics of the four guidelines are vague. North Korean public documents and statements regarding the Line of Self-Reliant Defence are filled with ideological justifications and praise for the defense planning doctrine as the leader's ingenious efforts to defend the nation, or claims over how the four guidelines are inherited from the KPRA and the Korean War.[3] Pyongyang also publicizes very little on the implementation progress, and detailed announcements and publications such as white papers, procurement reports, defense plans, and so forth simply do not exist. At best, the DPRK has published a range of speeches and essays by Kim Il-sung that talks about the importance of the Line of Self-Reliant Defence and the leader's delight in the progress of building the KPA in accordance with the doctrine. For example, at the 30th anniversary of the DPRK government on 9 September 1978, Kim Il-sung declared that the implementation of the Line of Self-Reliant Defence was complete, making the "KPA a revolutionary force and . . . [made] the state into an impregnable fortress."[4] Both Kim Jong-il and Kim Jong-un have also made references to the Line of Self-Reliant Defence, citing it as the bedrock of constructing a "powerful military guarantee."[5]

Despite the propagandistic descriptions by the regime, the Line of Self-Reliant Defence was not some ideological slogan but a genuine defense planning directive. The actual impact of the doctrine is not only evidenced by the developments in the KPA's capabilities since the 1960s, but also the significant political-economic reconfigurations to enhance and expand the military industry sector, as well as the regime's command and control over the armed forces. Rather, the secretive and vague nature of the Line of Self-Reliant Defence suggests that the specific plans are much more sophisticated, going beyond the confines of Pyongyang's official statements and only accessible to the most senior cadres involved in defense planning. Still, although the DPRK genuinely pursued the Line of Self-Reliant Defence, the process was far from smooth, revealing the cleavages between the state's interests and capacity. On the one hand, the DPRK is fixated to centralization, politicization, and inheritance to ensure the regime's legitimacy and survival. On the other hand, the state's limited economic and industrial capacity has set constraints on R&D, acquisitions, construction, and O&M. While defense planning dilemmas are faced by all governments, the DPRK demonstrates an extreme case. In the pursuit to achieve the Line of Self-Reliant Defence, the DPRK became enslaved by their rigid policies and self-inflicted capacity problems, compelling them to take on heavy burdens and make serious trade-offs between their interests for regime security and military readiness.

Cadre army

The "cadre army" guideline purports to enhance the KPA personnel's leadership and combat effectiveness through political loyalty and technical proficiency so that they can carry out the tasks of higher ranks.[6] Kim Il-sung had argued for the building of a cadre army since the mid-1950s to qualitatively alleviate the shortage of cadres that was seen by the leader as one of the reasons for the retreat during the Korean War.[7] However, while the guideline stresses the importance of sharpening the personnel's practical knowledge and skills, there is greater emphasis on political discipline and loyalty to the leadership.[8] As the KPA continued to grow in size, it became evermore important for Kim Il-sung to maintain absolute command and control, leading to the strengthening of the GPB and the establishment of WPK committees and cells in the various levels of the armed forces.[9] Thus, the "cadre army" guideline was an essential element to manage a quantitatively large armed forces that is well integrated under the leadership's command and control.

The emphasis on leadership and integrity based on ideological discipline, responsibility and ethics are the central themes to many of the literatures published by the GPB.[10] As Kim Il-sung strengthened his authority, the notion of military discipline began to emphasize much more on loyalty to the leadership and traditions inherited from Kim Il-sung's anti-Japanese partisan campaign.[11] The emphasis on the leadership's centralized and politicized command and control consequently molded the concept of military professionalism. In the KPA, all personnel must abide by the Ten-point Terms of Obedience issued on 30 November 1977 that includes: adherence to codes, adherence to socialist laws and orders, execution of military orders, execution of tasks ordered by the party and political organs, fluency in handling and management of weapons, protection of information, participation in both military and political training, affection for the people and not violating their property, protection of military and state assets, and unity and solidarity within the armed forces under good customs.[12] The Ten-point Terms of Obedience highlights the importance of not only discipline and professionalism from the military standpoint but also unwavering loyalty to the regime. While the former is essential to ensure practical competence, it is the latter that assumes higher priority. The DPRK's emphasis on ideological discipline essentially altered the way human resources are managed, giving the GPB greater authority. Hence, no matter how skilled or knowledgeable one may be in practical aspects, any blotches in political discipline or loyalty would affect their prospects for promotion and post-service careers, or worse, lead to severe punishment.

The intensive level of politicization significantly impacted the KPA's training philosophy and regimes. At the Tenth Plenary Meeting of the Fifth CCWPK in February 1975, Kim Il-sung laid out the "five-point policy" in strengthening the KPA personnel, consisting of: strong revolutionary spirit, skillful and clever tactics, "steel-like" physical strength, accurate marksmanship, and strong discipline.[13] Later in the 1990s, Kim Jong-il issued the "four-point training principles" based on: self-reliance, political ideology, combat, and science.[14] Over the years, various military publications frequently referred to the training policies and principles laid out by both Kim Il-sung and Kim Jong-il as the cornerstones of strengthening the KPA.[15]

Indeed, there are substantive aspects to the management of KPA, where officers are assigned to positions according to suitability. For instance, Kim Jong-il argued that field officers (below division and brigade levels) should be young because of their physical abilities, while older officers are more suitable as political officers for their discipline, experience, and knowledge.[16] Although the idea was probably inherited from Kim Il-sung, it proved to be more vital for Kim Jong-il as he assumed command of the KPA during the challenging times of the 1990s. In a meeting with KPA officers in 2005, Kim Jong-il admitted to the challenges in managing the KPA due to generational changes, arguing that the younger generations did not experience the hardships of the older generations.[17] Against this backdrop, Pyongyang focused on ensuring that the younger generations who grew up during the times of economic hardships remained disciplined and loyal to the traditions of the KPA and, of course, the leadership.

From the military readiness viewpoint, the centralized and politicized approach provides advantages in enabling the strict command and control over the armed forces – particularly in rules of engagement. The counterargument is that decentralized control has its benefits in providing greater flexibility at the tactical levels to achieve the state's strategic goals. Yet for the DPRK who depends on calculated brinkmanship and exploitation of "gray-zone" situations to exercise their military leverage, the decentralized approach would have its downsides, as any miscalculation, miscommunication, misinformation, misinterpretation, and misjudgment at the tactical level would lead to strategically catastrophic results.

The disadvantage of the "cadre army" principle is that the KPA's operations are often affected and meddled by politics. The KPA Party Committee system not only harmonizes but also checks-and-balances the field and political officers with the latter essentially holding greater authority. Technically, GPB commissars and officers hold the same rank or one rank below their GSD counterparts, but they have the authority to scrutinize any of the operational activities and plans such as composition of units, tactics, training regimes, and so forth.[18] Obviously, conflict of interest would occur. According to Scalapino and Lee, there were tensions between the GPB and GSD officers during the earlier years, although such problems were overcome with the establishment of the aforementioned KPA Party Committee system.[19] Still, the political commissars and officers have significant authority that allows them to alter or even reject plans drafted by commanding officers – even if they are essential from the operational or tactical standpoint. Of course, the DPRK does not disregard the importance of practical aspects, and it would be unreasonable to think that the WPK would completely reject anything that enhances the KPA's readiness and effectiveness. Nevertheless, the problem is not simply about politics, but the bureaucratic process that could undermine the innovative and pragmatic ideas from the field levels, essentially impacting the overall readiness and functional efficacy of the KPA.

Military modernization

The "modernization of the armed forces" guideline is about enhancing the KPA's technologies and techniques for warfare as well as the autonomous production of hardware. The guideline, however, was not necessarily about acquiring and utilizing new and powerful technologies, but rather about producing and skillfully utilizing assets to effectively disrupt and penetrate the vulnerabilities of technologically superior forces. The DPRK's unique conceptualization of military capabilities date back to Kim Il-sung's anti-Japanese guerilla partisan campaign, where he embraced the idea that technological shortfalls could be overcome by innovative tactics and techniques. As the KPA developed into a genuine military institution, Kim Il-sung acknowledged the importance of military technologies while also emphasizing that technological modernization and utilization should conform to particular conditions and styles.

The first and foremost aspect is the focus on capabilities that accord with the KPA's warfighting concepts and doctrines. The DPRK's "military strategy" is based on three pillars: "surprise attacks," "quick and decisive wars," and "mixed tactics."[20] Based on the military strategy, the KPA's "principles of war" consist of: "two-front war," "surprise," "mass and dispersion," "maneuverability," "initiative," "operational security," "annihilation," "combined operations," "mobility," and "rear area protection."[21] Finally, the "tactical doctrines" include: "sustainment;" "camouflage, concealment, cover, and deception;" "echelon forces;" "KPAAF and KPAN employment;" and "terrain appreciation."[22] Thus, by modernization, the DPRK seeks to get an array of assets, including: ballistic missiles for power projection; massed but fast and maneuverable units for asymmetric and swarm attacks; firepower for barrage attacks; and disruptive operations enabled by cyber and electronic warfare capabilities.

The second aspect is the emphasis on skills that draws on Kim Il-sung's concepts of guerilla warfare. Leading up to, and during the Korean War, Kim Il-sung gave a series of speeches on tactics and the importance of skills. Kim Il-sung's comments often stressed the importance of exploiting the geographical features of the Korean peninsula such as the mountains, rivers, and long coastlines, as well as the ability to operate in adverse weather and night.[23] The emphasis on skill and the human factor parallels Mao Zedong's ideas on "protracted war" that placed people ahead of weapons technologies.[24] Yet what differed in the DPRK's case is the extent to which Kim Il-sung continued to systemize the "mind over matter" mindset even as the KPA modernized, emphasizing the innovative use of infrastructures, materials, and platforms, rather than innovations in technology.[25] Despite the ideological explanations, the notion of doctrines over technologies is logical and is in fact an essential and pragmatic way to design and execute a variety of operations that effectively use assets.[26]

The third aspect, which relates to the second, is how the cocktail of regular and irregular warfare has made the KPA selective about the theater of wars, focusing on the areas where it will be most effective while avoiding the areas where it will be disadvantaged. The emphasis on ground warfare was much because it provided not only the set of capabilities that can take advantage of the geographical features of the Korean peninsula for asymmetric attacks and insertion of commandos but also the fact that the Seoul metropolitan area is within range of the artillery and MLRS units. At the same time, the DPRK has abstained from air and sea supremacy not simply because of the costs to acquire the capabilities but also because these domains are far less exploitable for asymmetric warfare and that the DPRK's strategic scope focuses on the Korean peninsula and the immediate periphery. Thus, with the exception of SLBM, the DPRK focused more on air and naval capabilities for: anti-access and area-denial; fire-support; transportation; intelligence, surveillance, target acquisition, and reconnaissance (ISTAR); as well as disruptive, guerilla attacks against enemy infrastructures, logistics, and forces within the confines of the Korean peninsula as opposed to air and maritime supremacy.

Naturally, the perceptions toward military modernization had significant impact on the DPRK's defense planning, although much was heavily politicized. In 1969, Kim Il-sung purged the then Minister of MNS Kim Chang-bong and other military hardliners for pursuing the acquisition of supersonic aircraft and direct-firing artillery instead of " 'howitzers or mortars' and 'low-speed, low-flying airplanes' better suited for the geographic conditions of Korea."[27] None of this, however, meant that Kim Il-sung was uninterested in advanced capabilities. Just months before the aforementioned purges, Kim Il-sung in fact gave a speech at the Kim Chaek University of Technology, where he stated, "at one time we told you to stop building helicopters and the like. But it is now high time to make helicopters and other types of aircraft and modern automatic weapons such as rockets in our country."[28] Going even further, Kim Il-sung in the mid-1960s talked about his desires to acquire power projection capabilities to strike the US or at least the US forces in East Asia, arguing that "if a war breaks out, the US and Japan will also be involved . . . to prevent their involvement, we have to be able to produce rockets which fly as far as Japan."[29] The seemingly inconsistent set of arguments and actions is not so much about the change of mind, but rather Kim Il-sung's efforts to dictate the defense planning process with zero-tolerance for lobbying.

There are also patterns in the DPRK's military modernization process. A key characteristic of the DPRK's military modernization program – particularly since the issuance of the Line of Self-Reliant Defense – is how it is based on a long-term schema that focuses on advancing the military industry's capacity for autonomous production, as opposed to a short-term schema of purchasing off-the-shelf for immediate operationalization. In most cases, the DPRK's military modernization has been a three-stage process that starts with acquiring (or on some rare occasions capturing) platforms from abroad, studying those technologies, and then producing customized variants with additional systems attached.

The one key advantage of the DPRK's military modernization program is the level of creativity attained despite the resource and technological shortfalls. Over the years, Pyongyang has taken a mix-and-match approach to produce a variety of platforms using the limited technologies they possess. On the ground, the

DPRK has mated various weapons systems and electronic equipment to some of its heavy-duty vehicles such as the VTT-323 to field an array of platforms for different purposes, including anti-air defense, amphibious operations, armored transportation, maneuverable combat, and so forth. In the sea, many of the surface combatants are built or retrofitted to install a variety of anti-ship and anti-submarine weapons systems. Even regarding aircraft, the DPRK has retrofitted some of its dated aircraft for asymmetric applications. Such is the case of the low-flying 1940s An-2 turbo-prop biplane that has been modified with canvass and wooden material to leave less radar signatures and also configured to launch the KN-01 anti-ship cruise missile (ASCM) and other payloads.[30] Indeed, the downside is that the aforementioned retrofitting measures would not be next-generation advancements. Moreover, there are bound to be limitations in the compatibility of certain systems (e.g., C4ISTAR), and some platforms such as jet aircraft would be less forgiving to retrofits. Nevertheless, despite some of the constraints in the mix-and-match approach, they have been sufficient in fielding a variety of capabilities for asymmetric effect.

Another key advantage of the DPRK's style of military modernization is the level of restraint in committing to excessive investments. Indeed, the prioritization of developing the military sector over state economic development is morally questionable, and the push for self-reliant ways of military modernization was expensive. Yet still, the DPRK pursued its self-reliant military modernization in the most cost-effective form possible, with careful planning that takes into consideration of the state's limited economic and industrial capacity. Even if the DPRK idolized the weapons in the Soviet and Chinese militaries, it also understood that it neither has the capacity to develop nor operate such kind of forces. Pyongyang could not afford to make ambitious acquisitions, as doing so would lead to excessive investment and O&M costs, a capability surplus that undermines cost-effectiveness, as well as ebbing the KPA's asymmetric strengths. The DPRK, therefore, focused on the capabilities that are most vital and cost-effective. Thus even though the pursuit of ballistic missiles and WMD, as well as cyber and electronic warfare capabilities have been undoubtedly expensive, they were perceived as more cost-effective than qualitatively overhauling the whole KPA.

One key disadvantage is that the focus on indigenous production despite limited capacity inevitably forces the DPRK to be constantly behind the technological curve. Although the DPRK was careful not to become over-dependent on its benefactors, it was still – and continues to be – dependent on acquiring technological templates. Such problems reveal how the DPRK's military technologies are confined to the technologies accessed and acquired, and true forms of technological innovation are still limited. Moreover, as discussed in the previous chapter, the DPRK's access to technology has been far from consistent – particularly in conventional capabilities. Even though the various military parades in the DPRK often showcase new platforms that evidence some levels of modernization, the actual technologies themselves are certainly not the best in current-day terms – particularly in the case of C4ISTAR systems. In naval platforms, modernization has taken place with the construction and commissioning of new vessels, but still fall short of the capabilities required to be effective. In air capabilities, advancements in aircraft have been minimal since the 1990s, and more attention has been paid to the upgrading of some anti-air missiles. Even for the ballistic and cruise missiles where advancements have been most evident, there are still questions concerning accuracy, range, survivability, and so forth. Hence, although the DPRK has worked to modernize the KPA and the mix-and-match approach was effective to some extent, there are still lags in the level of technological advancement achieved.

The other major disadvantage is the very size and weight of the armed forces that constrain modernization. The advantages of the DPRK's quantitative approach to defense planning depend much on whether it can sufficiently produce and distribute the weapons systems to the whole armed forces. But given the capacity problems in the military industry, the DPRK is constrained in adequately modernizing the KPA across the board. Under such circumstances, two options are available. The first option would be to concentrate modernization to the capabilities and units most vital to the KPA. Yet this would cause severe imbalances with the rest of the forces, undermining the ability to effectively and efficiently coordinate and integrate the capabilities for large-scale operations. The second option would be to forgo modernization and innovatively use the existing capabilities with guerilla tactics to save investment costs. That said, the option would not only cause upticks in the O&M costs to keep the capabilities in combat-worthy condition but also denies avenues to acquire capabilities that are nonetheless essential to engage and deter the US, ROK, and Japanese forces (e.g., strategic weapons, cyber and electronic warfare systems, and anti-access and area-denial capabilities).

In sum, Pyongyang's approach to military modernization has been the most cost-effective means to attain capabilities that match with its warfighting concepts and doctrines while also staying within the boundaries of its political and economic constraints. One could argue that Kim Il-sung and his successors followed their own Pareto's principle to ensure that the KPA has the capabilities sufficient to effectively carry out hybrid warfare. While true to some extent, Pyongyang's approach also resulted in the assortment of capabilities with varying levels of readiness that make it hard to create an effective and efficient system of readiness. Indeed, the acquisition of ballistic missiles in recent decades has certainly compensated for the gaps in firepower and power projection. Still, such measures are merely band-aid solutions, and the DPRK will nevertheless face tough questions on what capabilities are needed to fill the technological deficit.

Arming the citizenry

The "arming the citizenry" guideline is about the massed augmentation of military and paramilitary forces with farmers and workers armed under the slogan of "a gun in one hand, and a hammer and sickle in the other" so that the whole society sustains a high level of readiness for a "people's war." Yet although Kim Il-sung's statements on the guideline seem to focus on reserve units for homeland defense and logistical roles, it is closely tied to the need to possess a quantitatively strong active component that is enabled by a robust military service system. The DPRK's mass-enlistment of personnel not only allows the KPA to sustain a large standing force but also enhances the readiness of the reserve forces that are largely composed of those who previously served active duty.

Article 86 of the current constitution states that "National defence is the supreme duty and honour of citizens. Citizens shall defend the country and serve in the armed forces as required by law." Moreover, the DPRK does not hide that the "duty and honor" is about defending the leadership who serves to uphold the "sovereignty of peasants and workers" and "socialism."[31] Kim Il-sung's move to boost the active and reserve forces was initially motivated by his view that the chronic personnel shortages and weaknesses in local-level militias was one of the major problems in the DPRK's efforts during the Korean War.[32] The pursuit to build a massed military accelerated in the immediate post-armistice period with the withdrawal of Chinese troops from the DPRK. Attaining a quantitative edge in personnel was difficult, given that the population of the North was smaller than that of the South. Nevertheless, as the technological gaps with the US and ROK counterparts became increasingly apparent, Kim Il-sung viewed that in the short-term, the militarization of the masses would be more economical and politically beneficial than upgrading the KPA's technologies.[33]

Generally, conscripts begin their service upon graduating from senior middle school. One notable aspect of the DPRK conscription system is the duration of military service that has been significantly extended since the 1950s. Currently, the length of obligatory service is generally 10 to 12 years for men (13 years or indefinite for special operations units) and roughly 5 to 7 years for women. In most cases, male enlistees reach the rank equivalent of Master Sergeant or Sergeant Major by the end of their mandatory term. As for female enlistees, they generally serve in support units such as in logistics, communications, medics, while some are assigned to anti-air artillery units. Moreover, depending on their competence, specialty, and political profile, soldiers have the opportunity to either be commissioned or stay on as tenured non-commissioned personnel.[34] Even for those discharged, most of them serve as reservists, enabling the reserve components to attain some level of readiness.

Selection criteria for officers are strict, based on not only competence but also political profile. Generally, commissioned officers are prior-enlisted, specialists, or those who enter directly from officer candidate schools.[35] Naturally, the criteria for officers in specialized and technical positions such as aviators, submariners are stricter, requiring not only high level of political credentials but also higher levels of physical and technical aptitude.

Regarding the reserve forces, Pyongyang systemized the WPRG in 1958, the RMTU in 1963, and the RYG in 1970. Each of the reserve forces differs in nature. The RMTU is the ready-reserve unit made up of men between the age of 17 and 50 and unmarried women between the age of 17 and 30.[36] The WPRG is made up of men between the age of 17 and 60 and women between the age of 17 and 30 who are not enlisted in the RMTU or in other paramilitary services.[37]

The RYG consists of senior middle school students between the ages of 14 and 16 and their roles focus on political activities and preparatory training for their future service in the KPA.[38]

As for the paramilitary forces, the largest is the KPISF affiliated to the MSoS and is akin to the People's Armed Police in China. The KPISF comes under the command of the MND in wartime for either combat mobilization or homeland security. In addition, the KPA Guard Command is a massive unit that could be categorized as a special operations corps, but their primary role is the personal security of the leadership and thus should be distinguished from the rest of the KPA that is mobilized for broader military operations.

The introduction of mass recruitment policies combined with the increase in the population not only led to significant boost in personnel but also created an ecosystem of readiness between the active and reserve components. The large active and reserve components allow smoother personnel management for the DPRK. For instance, the RYG readies the members into their possible future tenure in the KPA or at least in the RMTU and WPRG. At the same time, much of the members of the WPRG and also the RMTU have served in the KPA, allowing the reserve component to maintain a certain level of readiness. Moreover, members of the RMTU with good credentials also have the opportunity to be commissioned as officers in the KPA. Even for roles, the active component focuses on defense, while the reserves focus on civil defense, logistics, and homeland security. Thus, the active forces are dependent on the capabilities of the reserve forces and vice versa.

The key advantage of both the mass-enlistment and reserve militia system is that it allowed the DPRK to kill two birds with one stone, in what Scalapino and Lee correctly described as a system that "unites the political and military objectives of the regime."[39] Militarily, the DPRK sought to attain a high number of personnel to quantitatively beef up both the active and reserve forces for not only combat and logistical roles but also labor. Politically, both the mass recruitment system served the DPRK well. In the DPRK, the word "mujang" (armament) is not simply about technical combat readiness but also about political discipline – "sasangjeok mujang" (ideological armament). The "arming the populace" guideline, therefore, allowed the regime to politically control a wide cross-section of the populace through the military, consequently leaving little room for organization of political alternatives by providing the regime with another layer of surveillance over the society as well as an additional channel to propagate their policies and gain local support.

The major disadvantage of arming the masses is the greater exposure of both the active and reserve forces to the state's limited capacity and socio-economic problems. As the KPA grew into a force with emphasis on human capital, its very size became a major burden and liability. Given the economic constraints, the DPRK struggles to provide basic supplies to its soldiers let alone qualitative improvement. Consequently, the readiness deficiencies became increasingly apparent, where some units are better equipped and trained while others are marginalized and suffer from deteriorating state of health and morale. Such problems consequently undermine the purpose of having a massed military, where the KPA cannot count on the whole forces to be mission-ready.[40]

The mass armament guideline certainly enabled the KPA to mount a massive force while also serving as a mechanism to control the populace. At the same time, the policy also compelled the DPRK to take on greater burdens to effectively operate and sustain the readiness of millions of active and reserve personnel. Therefore, the mass armament guideline granted the KPA's quantitative strengths while also causing the qualitative weaknesses.

Nationwide fortification

The "fortification of the state" guideline purports to garrison key military, industrial, and political installations. The rationale for the guideline is four-fold: first, to protect the DPRK's vital military assets from enemy strikes, airborne and amphibious assaults; second, to conceal critical assets from enemy ISTAR capabilities; third, to facilitate the KPA's counter- and second-strike capabilities; and fourth, to ensure that the key industries and logistical supply chain can adequately function under war conditions. The fortification guideline is based on the lessons drawn from the Korean War when the KPA suffered extensive damage from aerial bombardments but further embodied as airpower and missiles became evermore critical components of modern warfare.

The DPRK already began to construct underground bases and factories during the Korean War, but the quantity and quality of its underground facilities grew significantly in the 1960s when Pyongyang devoted significant resources to finance the fortification project. At the Fifth WPK Congress in November 1970, Kim Il-sung declared that the key military and industrial installations had been "fortified."[41] The result was the construction of a cohort of: fortified ground, naval, and air positions and strongpoints; ordnance factories and logistical bases that are fully or partially inside mountains; tunnel networks; as well as the concentrated deployment of artillery, MLRS, and anti-air and anti-ship defense systems in strategically vital areas such as Pyongyang, MDL, and coastlines.

Today, reports estimate that the DPRK has constructed over 10,000 underground facilities and factories, including an array of hardened artillery sites (HARTS) positioned below the Pyongyang–Wonsan Line in mountainous terrains, along the coast, and remote islands.[42] Many of the HARTS are constructed into "E," "I," "U," and "Y"-shaped tunnels not only to provide multiple options for fire but also to serve as key lines of communication.[43] Furthermore, fortification also involved strategically locating and constructing the installations, where many of the KPA bases are located near mountains connected to the HARTS and other underground berthing facilities.[44]

Another aspect of the fortification guideline is the flexible use of infrastructures. The DPRK has constructed bases with the expectation that some of them will be critically damaged in wartime, in which case the KPA will utilize auxiliary or improvised infrastructures for military purposes. While the auxiliary or improvised infrastructures are in substandard condition, the KPA is tasked to bring them into operational levels when orders for wartime readiness are handed down.

The advantage of the fortification guideline is that the DPRK has well exploited its terrain in ways that benefit the KPA's defensive and offensive operations. In defense, the underground installations and tunnels provide protection of assets from enemy firepower. As for offense, the KPA can utilize the HARTS and tunnels to conduct first- and/or second-strike with little warning.[45] Moreover, the fortified installations also provide a high degree of concealment, creating enormous ISTAR and logistical challenges for the opponents to locate, track, and strike the sites. The US and ROK have only identified some of the KPA's underground installations and tunnels, and the actual number and their locality remain unknown. Consequently, the logistical challenges to carry out an attack would be high, particularly given the uncertainty over the locality, structure, and number of these underground installations. Such problems would lead to a lethal "whac-a-mole" scenario where an attack on the selected installations may only lead to an attack from another.

The other advantage is the availability of various infrastructures. The KPA platforms are modified, and the personnel are trained to operate on alternative roads, runways, ports, and tunnels.[46] Although the level of effectiveness would be much inferior to that of proper infrastructures, such options do allow the KPA to have some flexibility in operations and logistical support. Even in defense, the KPA is indoctrinated to form various types of blockades on land as well as anti-air and anti-ship measures to disrupt and slow incoming forces.

From the cost-benefit viewpoint, the nationwide fortification project presents more disadvantages than advantages. Naturally, the economic impact of the fortification guideline was devastating. According to estimates by the Central Intelligence Agency (CIA), the construction of underground military installations costs at least three to four times than standard facilities as well as taking much more time.[47] Given that Pyongyang embarked on the mass fortification project in the early 1960s, there is little doubt that it contributed to the economic decline in the following years. Yet the big question is about cost-effectiveness, given the dubious benefits even in military terms. While the underground installations and tunnels provide the KPA with some concealment and protection, the mobility and flow of units and personnel, as well as weapons production and maintenance are compromised due to narrower space and ventilation issues. For instance, regarding the underground airbases such as those in Jangjin, Onchon, and Wonsan, Bermudez notes the longer takeoff intervals due to the rapid buildup of exhaust gases, and that direct landings "were generally deemed as impractical."[48]

Another disadvantage is that the developments in technologies are denying the effectiveness of fortification. In the age of advanced C4ISTAR capabilities, the KPA has to ensure that any movements are absolutely discreet and minimal to avoid its underground facilities from being identified, literally illustrating the old saw – "you can hide, but you can't run." Even with regards to the protection against bombardments, advancements in munitions are chipping away at the effectiveness of bunkered facilities. Although the DPRK strengthened its underground military installations with reinforced concrete in the 1990s, advancements in cruise missiles and "bunker buster" munitions such as the GBU-28 or the GBU-57A/B Massive Ordnance Penetrator would destruct or at least severely disrupt the KPA's underground facilities.

In hindsight, nationwide fortification proved to be plausible given that the DPRK's vulnerability to enemy strikes became much greater than the times of the Korean War, with the advancements in ballistic and cruise missiles, and "bunker buster" munitions. Thus, the concealment and protection of military assets and facilitation of one-off surprise attacks do benefit the KPA. At the same time, the construction of fortified infrastructures proved to be extremely costly. Obviously, there was zero-sum balance with the civilian economy. But even from the defense planning viewpoint, the excessive devotion of resources to the construction and operation of fortified infrastructures have undermined Pyongyang's capacity to strengthen the other critical areas of KPA's readiness. Despite the costs and diminishing effectiveness of the project, reconstructing the installations or constructing alternatives is also extremely difficult. Shifting away from underground installations would not only increase the vulnerability of the KPA but also implicate high costs and time to construct new infrastructures. While the DPRK could indeed construct new installations with better specifications, they will be more or less in line with the infrastructures already built under the fortification guideline.

Macro-level defense planning dilemmas

For the DPRK leadership, the Line of Self-Reliant Defence was a catch-all solution to enhance and manage the KPA's readiness while ensuring the regime's political security. Yet the Line of Self-Reliant Defence was overambitious and formulating the right scheme to achieve maximum levels of both political control and military readiness were far beyond the state's capacity. Consequently, the divergence between interests and capacity created dilemmas between the focus on regime security or military readiness. As given in Table 5.1, the two options present a polar set of advantages and disadvantages concerning the guidelines of the Line of Self-Reliant Defence. On the one hand, focusing on regime security would entail measures focusing on centralization and politicization that compromises measures to genuinely enhance military readiness. On the other hand, greater focus on military readiness would require reconfigurations for more cost-effective measures that necessitate moderations to the centralized and politicized management of the KPA.

Realistically, remedying the shortfalls of the Line of Self-Reliant Defence is impossible without overhauling Pyongyang's defense planning. Yet embarking on a more balanced approach would unearth major decision-making dilemmas between the interests and capacity of the regime.

Above all, there is the political factor. Fixing the long-standing problems and formulating innovative solutions would require a process that is less centralized and politicized while allowing greater input from technocrats. Although the

Table 5.1 Decision-making dilemmas based on the Line of Self-Reliant Defence

Focus	*Options*			
	Regime security		*Military readiness*	
	Approach	*Shortfalls*	*Approach*	*Shortfalls*
Cadre army	Emphasize integrity and loyalty of personnel to the regime	Practical aptitude of personnel for military operations	Prioritize and emphasize practical aptitude	Lower levels of centralized and politicized command and control
Military modernization	Massed-armament of indigenously produced weapons	Lacks in high-end capabilities	Procurement of high-end capabilities	High investment and O&M costs as well as major force structural adjustments
Arming the citizenry	Maintain current regime for political control and mobilization	Lack of provisions and vulnerability to socio-economic issues	Focus on mobilization for national defense by improving provisions and training regimes	Higher costs for provisions and improvements in training regimes
Nationwide fortification	Maintain current infrastructures	Lacks in effectiveness and efficiencies for defense and mobilization	Reconstruction for more effective and efficient means of defense and mobilization	Costs for reconfiguration and reconstruction
Political-economic risks	Lower		Higher	
Cost-effectiveness	Lower		Higher	
Overall advantages	Political control and lower costs		Military readiness	
Overall disadvantages	Military readiness		Political control and higher costs	

dynamics of the leadership is different, China presents an interesting case, where the People's Liberation Army has become more "professional and technocratic," lubricating the military modernization while also keeping the party-military relations "symbiotic."[49] However, the question is the degree of change that is required to facilitate greater developments that focus on military readiness. Since the dawn of the regime, Kim Il-sung and his successors constructed an architecture whereby the leadership has outright control over the management of the KPA. Moreover, the Line of Self-Reliant Defence was a grand project that came at the expense of state development as well as shaping the functionality and identity of the state. Revising or even attempting to hedge would be risky, as this would unearth contradictions that undermine the defense planning process and the regime's legitimacy.

Economically, despite the establishment of, and heavy devotion of resources to the indigenous military-industrial complex, Pyongyang failed to build substantive capacity to sharpen the KPA's readiness. As discussed in Chapter 4, while the DPRK indeed established its own military-industrial complex, the actual capacity has been limited and the access to technologies has been inconsistent. Simply put, the Line of Self-Reliant Defence lacked economic considerations where the resource and technological demands were far beyond the DPRK's capacity. One can also blame how the resources were spent. For instance, the DPRK could have devoted more resources to acquire high-end weapons systems than investing in fortified bases and building a massed force structure. At the same time, overhauling the output of the Line of Self-Reliant Defence would prove to be extremely expensive, as this would not only require investments in new hardware but also reconfiguration of O&M procedures. Hence, the Line of Self-Reliant Defence stretched the limits of the DPRK economy, making it hard to make any further sacrifices to strengthen the KPA's readiness.

Revising the Line of Self-Reliant Defence is also difficult given how the four doctrines are interlinked and intertwined. For instance, the "cadre army" and "arming the citizenry" guidelines are linked to enable smoother and better integrated wartime command and control over the active, reserve, and paramilitary forces.[50] The "military modernization" guideline is also linked to the other three to ensure that the KPA's capabilities: conform to the concepts and doctrines of the leadership; are quantitatively sufficient to arm the massed forces; and operate well with the underground facilities for concealment, protection, and surprise attacks. The nexuses reveal that the Line of Self-Reliant Defence is not a collection of four independent guidelines but a community of interconnected guidelines. Although the synergy of the four guidelines undoubtedly shaped the armed forces according to the leadership's visions, the linkages are one of the very reasons in why the Line of Self-Reliant Defence is difficult to revise, as revising one guideline would inevitably affect the other with little alternatives to fill the gaps, consequently corroding the very architecture of the Line of Self-Reliant Defence.

For the DPRK, the political-economic switching costs and risks in altering its defense planning are far too high. At the same time, maintaining the current trajectory also has major consequences. The combination of the DPRK's obsession with regime survival and limited political-economic capacity leave little room to innovatively enhance the military's readiness, pushing the defense planning system into a high degree of path-dependence. Against this backdrop, the DPRK has to make major trade-offs to strengthen the KPA with minimal costs and disruptions. Thus, the biggest dilemma for the DPRK is which of the two consequences it is prepared to face first or willing to compromise – command and control or military readiness.

From the way in which the defense planning doctrine was implemented, it is clear that Pyongyang placed greater emphasis on the command and control over the KPA, consequently creating disadvantages in military readiness. Pyongyang's option was due to not only its obsession with regime survival but also the perception that the military readiness disadvantages are manageable and could be overcome as long as the KPA remains firmly under the leadership's command and control. The DPRK embarked on sharpening the KPA's readiness in ways that are economically feasible but do not compromise the regime's centralized and politicized command and control. This more balanced approach in pursuing the Line of Self-Reliant Defence by modernizing the KPA without compromising command and control has proved to be costly, but it was nonetheless the only way to achieve the state's strategic, military, and political ends.

Consistency despite change

To this day, the Line of Self-Reliant Defence has remained to be the bedrock of Pyongyang's defense planning. Yet despite the high level of consistency, the DPRK has periodically sought ways to strengthen the KPA without making any "switches" in its defense planning or compromises in its political and economic policies.

The first phase took place during the 1960s when Kim Il-sung introduced the Line of Self-Reliant Defence as the DPRK's defense planning doctrine. Prior to the issuance of the doctrine, the KPA was modest in size, with approximately 338,000 active personnel.[51] Regarding inventory, the KPA at this time was equipped with aircraft, artillery, vessels, and tanks from China and the USSR, and also started to build the capacity to produce its own chemical and biological weapons, and missile systems. Nevertheless, given the combination of internal and external uncertainties, Pyongyang felt that it needed to build its indigenous capacity to enhance the KPA's readiness, leading to the issuance of the Line of Self-Reliant Defence and opening the period that would be devoted to refurbishing the political-economic framework and constructing the industrial capacity to develop the KPA's massed, fortified, hybrid warfare capabilities.

One key agenda was to build the military–industrial complex. Nevertheless, the DPRK was dependent on both China and the USSR for major technological acquisitions in the short term. During the 1960s, the DPRK was able to gain some essential capabilities such as aircraft, light surface vessels, submarines, and ground vehicles that would not only be fielded by the KPA but also became the technological templates for some of the platforms the DPRK will produce in the future. Yet as explained in the previous chapter, Pyongyang's arms acquisitions from the 1960s onward hinged on the state of relations with Beijing and Moscow, leading to a mixed composition of Soviet originals and Chinese variants. Despite the occasional souring of ties, the USSR continued to be the most preferred and vital suppliers of critical technologies in not only conventional platforms but also missile and nuclear technologies.[52] Thus, for the KPA, the DPRK's rekindled relations with the USSR served as a blessing given that Soviet military hardware were more modern and authentic than the Chinese variants.

Developments were also evident in the KPA's force structure, with the establishment of various divisions and brigades to distribute its forces to strategically vital positions while also preparing them for the new capabilities to be operationalized in the coming years.[53] By the 1970s, the KPA consisted of approximately 401,000 active personnel, with a greater collection of mechanized units, vessels, and tactical aircraft.[54] Thus, although the actual capabilities of the KPA were still nascent, it was beginning to move toward a more massed, mechanized, and forward-deployed force.

Despite the economic burdens and geopolitical disruptions, Kim Il-sung was pleased with the progress, seen in his speech at the Fifth WPK Congress in November 1970 where he stated that the DPRK has "established an all-people and all-nation defense system" that has strengthened the KPA, armed the citizens, and built fortified infrastructures for self-defense.[55] Of course, not everything was achieved, but the foundations to develop the KPA in Juche-style were in place.

The second phase came in the 1970s and 1980s when the DPRK began to reap the benefits of the military-centric industrialization, facilitating greater self-sufficiency in armaments and growth in the KPA inventory. By the 1970s, the DPRK's military industry was able to manufacture more sophisticated platforms such as tanks, armored vehicles, self-propelled and towed artillery, MLRS, as well as some surface vessels and submarines. Developments were seen in aircraft, where the DPRK was building the capacity to produce the An-2, Il-28, Mi-2, MiG-15, MiG-17, MiG-21, and Yak-18.[56] By the late 1970s, the DPRK boasted that the KPA was "armed with the latest military technology and armaments to match [today's] modern warfare."[57] Yet despite the DPRK's efforts to indigenously produce a wider range of weaponry, Pyongyang was still technologically behind the curve. At most, the DPRK was only capable of producing variants or retrofits of platforms previously acquired from the USSR and China. Shortfalls were most severe in air and naval platforms. When relations with the USSR had rekindled in the mid-1980s, the DPRK was able to acquire new tactical aircraft such as the MiG-23, MiG-29, and the Su-25.[58] Still, the acquisitions did not lead to any significant technological developments in the DPRK's arms industry, and Pyongyang seemed to have placed greater emphasis on repairing and upgrading old platforms to quantitatively arm the KPA.

Progress in the strategic weapons project were incremental and still in the early stages during the 1970s and 1980s. The DPRK acquired two SCUD-B missiles from Egypt in the late 1970s in return for the KPA's participation in the Yom Kippur War. Pyongyang then began to produce its own tactical ballistic missiles such as the Hwasong-5 SRBM based on the SCUD-B and the Hwasong-6 based on the SCUD-C. The first test launch of ballistic missiles took place in April and September 1984 when the DPRK fired a total of six Hwasong-5 from the Musudan-ri Launch Facility in North Hamgyong Province toward the Sea of Japan (Korean name: East Sea), albeit half of them reportedly failed.[59] Although the DPRK's missile capabilities were still nascent, they would nevertheless serve as a technological base for future developments.

Force structural developments were notable with the number of active personnel reaching approximately 782,000 in 1980.[60] The KPA now had a greater pool of personnel at its disposal for both military and ancillary roles. While the former was vital to enhance the KPA's readiness, the latter gained greater importance given the state's economic decline, requiring much human capital for construction, factory, and farming labor. In combat roles, the greater number of personnel and weaponry granted Pyongyang with the capacity to thicken the forward-deployed ORBAT of the KPA below the Pyongyang–Wonsan line.[61] Thus, by the end of the 1980s, the foundations of the quantitatively dense, forward-deployed KPA that we see today were taking shape.

Although there were notable developments in the KPA during the 1970s and 1980s, capacity problems inevitably limited the pace and volume of the developments. Combined with the slowing economic growth, the DPRK was now in a tenuous position, and the technological gaps with the US, ROK, and Japan were becoming more apparent. Moreover, the DPRK also witnessed new developments in modern warfare, with technologies such as unmanned systems, as well as cyber and electronic warfare capabilities becoming a norm. Reorienting to the new realities was essential, yet given the capacity and circumstances, the KPA was as ready as they could be.

The third phase took place from the early 1990s when Kim Jong-il assumed his roles as the SCAF and Chairman of the NDC. In 1992, the DPRK amended its constitution, incorporating the Line of Self-Reliant Defence as the fundamentals of the state's defense planning. Geopolitically, however, circumstances were becoming increasingly unfavorable for Pyongyang. Both Beijing and Moscow became less generous in providing weapons and advisors, and the collapse of the USSR not only restricted the DPRK's opportunities to modernize the aging KPA inventory but also meant the loss of a key economic and security guarantor. Furthermore, Pyongyang's poor economic management dragged the DPRK into negative growth that will last for nearly a decade. Although Kim Jong-il in 1993 declared a "semi-war" situation, the increasing demands to enhance the KPA's readiness despite diminishing capacity further deepened the DPRK's defense planning dilemmas.

By the 1990s, the number of active personnel in the KPA had reached 1.1 million.[62] At the same time, the effects of the political-economic troubles on the KPA inventory were becoming evident. Not only was Pyongyang struggling to acquire new technologies from abroad but its military industry was also struggling to produce more modern weaponry. While the DPRK had collected some platforms from China and the USSR, the ability to autonomously produce more modern platforms – particularly air and naval – were limited. At best, the DPRK only had the capacity to make modifications to the Cold War platforms acquired from the benefactors, and modernization was essentially limited to artillery, MLRS, tanks, and various other armored vehicles, but these systems too were behind by 1990s standards.

Clear progress, however, was taking place in the strategic weapons program that was further prioritized to circumvent the shortfalls in conventional inventory. By the early 1990s, the DPRK was capable of producing the KN-02 SRBM which was based on the Soviet 9K79 (SS-21 Scarab) acquired through Syria.[63] While the KN-02 only has a range of 160 km, the mobile launch system made it suitable for tactical attacks on the battlefield as well as strikes on Seoul. Moreover, the DPRK also finished the production of a MRBM known as Hwasong-7 (or Nodong) that was test launched toward the Sea of Japan (Korean name: East Sea) on 29 May 1993. Developments were also seen in the much larger IRBM technology demonstrator known as Taepodong-1 that was fired over the Japanese archipelago on 31 August 1998. Although Pyongyang claimed that the launch was to put the Kwangmyongsong-1 satellite into orbit, it was widely believed either that the satellite launch failed or that it was in fact the test of a multi-stage, long-range missile.[64]

Developments were also seen in capabilities for cyber and electronic warfare which is packaged by the DPRK as "electronic intelligence warfare" (EIW). While the DPRK had always embraced the importance of electronic warfare since the formative years, it was alarmed by the technologies used in the US-led coalition's Operation Desert Storm in 1991. Realizing the new norms in modern warfare, Pyongyang scrambled to upgrade its EIW capabilities by establishing education and research institutions as well as working to install a variety of C4ISTAR assets. As for cyber warfare, the growing global use of the internet presented new opportunities for the DPRK to conduct attacks on not only enemy infrastructures and networks but also counterstability operations. While the EIW capabilities in the 1990s were still nascent, developments accelerated as time passed, becoming one of the DPRK's key assets.

Despite Kim Jong-il's efforts, the KPA's readiness was humiliatingly proven wrong with the defeats in the three naval battles in the Yellow Sea in June 1999, June 2002, and November 2009. Acknowledging that filling the technological gaps in conventional forces was insurmountable in the near-term due to economic difficulties, the DPRK pushed harder to develop its asymmetric and strategic weapons capabilities.[65] Yet the ballistic missile tests (or rocket launches) in the 2000s showed little significant progress. Developments, however, were taking place with the nuclear weapons program. In February 2005, the DPRK publicly declared its possession of a nuclear deterrent and then conducted its first underground nuclear test on 9 October 2006 and another on 25 May 2009. Indeed, the level of threat was still hypothetical due to the number of technological hurdles such as the miniaturization and reentry technology of the nuclear warheads. Nevertheless, it was evident that the DPRK had no intentions of turning back on its nuclear weapons program.

The fourth phase began from around 2010 which saw renewed efforts to revamp the KPA's readiness. The early part of this phase was marked by actions taken by the DPRK. In January 2010, Kim Jong-il along with cadres from the NDC, CCWPK, and KPA inspected the Combined Maneuvers exercises that promoted the KPA's readiness for cross-domain operations.[66] The significance of the exercises lied in the emphasis on conventional units in the ground, naval, and aerial domains. Within months, the KPA's conventional capabilities were mobilized against the South, with the sinking of the ROK Navy corvette Cheonan in March 2010 and the bombardment of Yeonpyeong Island in November. The Combined Maneuvers exercises and the two armed attacks in 2010 were clear indications that the DPRK was willing to strengthen its military leverage by using the KPA's conventional forces to penetrate the vulnerabilities of not only the ROK but also the US Forces Korea.

The death of Kim Jong-il on 17 December 2011 did not slow the momentum to strengthen the KPA's readiness. Rather, Kim Jong-un renewed the emphasis on modernizing both the conventional and strategic weapons capabilities of the KPA. At the plenary session of the CCWPK on 31 March 2013, Kim Jong-un issued the Byungjin line calling for the parallel development of the economy and nuclear weapons. In many ways, the policy is akin to Kim Il-sung's "parallel development of the military and economy" policy of the 1960s but this time with specific focus on nuclear weapons. The Byungjin line demonstrated Kim Jong-un's strong degree of confidence to rest his credibility on the strategic weapons program. Kim Jong-un's policy was real, moving on from technological demonstrators to a diverse range of vehicle-launch ballistic missiles including SLBMs. Even in tests, the number of ballistic missile tests since 2013 has far surpassed those conducted during the Kim Il-sung and Kim Jong-il eras combined and with a higher success rate. In 2018, the DPRK declared a moratorium on underground nuclear tests and ICBM launches, as well as the closure of the Punggye-ri nuclear test site. In part, the move was diplomatic, to demonstrate its part in reducing tensions while also setting the narrative in its interactions with the US and ROK. Yet Pyongyang's measures also indicate their confidence and satisfaction with their technological attainments, therefore focusing more on production rather than tests.

Developments in conventional capabilities have also been evident since the 2010s and into the 2020s with many of them showcased at the 75th anniversary of the WPK parade in October 2020 along with the new cohort of ballistic missiles. Over the past decade, the DPRK has acquired: range of armored vehicles; artillery and MLRS; submarines capable of firing SLBMs; frigates and surface effect ships; unmanned systems; anti-air/ship/tank missiles; cyber and electronic warfare capabilities; and others including some C4ISTAR systems. Developments are also seen in equipment, with greater levels of computerization that reveal some advancements in the KPA's C4ISTAR systems. In addition, upgrades to military infrastructures are also evident, particularly in missile-related installations and naval bases that indicate greater production capacity. Such developments are not simply to diversify and enhance the KPA's capabilities but also to create a system of assets to form a readiness kill-chain.

The KPA's modernization since the 2010s were based on a number of factors. The obvious one is technological, where upgrades to the KPA conventional inventory were long overdue, and the strategic weapons program had to advance for the DPRK to genuinely strengthen the KPA and also its leverage against the US, ROK, and Japan. Yet there are also internal factors. Politically, modernization of the KPA was vital for Kim Jong-un to prove that his policies to strengthen the armed forces under the Byungjin line were successful and worthwhile to demonstrate his credibility as the leader – particularly following the disastrous years under Kim Jong-il. Moreover, another essential domestic factor is the combination of the demographic trends and issues in health and welfare that undermine both the active and reserve forces, thereby requiring modernization to fill the gaps and sharpen the readiness of the armed forces.

The four phases of reconfiguration in the DPRK's defense planning were not improvised or chaotic but were planned and demand-driven, taking place during periods that were arguably the most challenging for Pyongyang since the Korean War. External factors such as the US presence in East Asia, strengthening capabilities of the ROK and Japan, and questionable relations with China and the USSR/Russia were certainly key factors that drove the DPRK to strengthen its military leverage. The internal factors were also significant, where Pyongyang's diminishing capacity and political uncertainties including leadership transitions led to significant emphasis on military affairs to legitimize and sustain the regime.

Yet although the four phases do reflect modifications in how the DPRK strengthened the KPA, there was a great deal of consistency in Pyongyang's defense planning. To this day, the Line of Self-Reliant Defence remains to be enshrined in the constitution as the foundational defense planning doctrine. Interestingly, when the Line of Self-Reliant Defence was codified into the constitution in 1992, the four guidelines of the doctrine were listed in a different order from the past. The amended constitution now stated the four guidelines as "train the army into a cadre army, military modernization, arm the entire people, and fortify the country," a change from the original order of "arm the entire people, fortify the country, train the army into a cadre army, and military modernization." The reshuffle reflected Pyongyang's view that the "arming the populace" and "fortification of the state" had been accomplished to a satisfactory extent while the guidelines of "cadre army" and "military modernization" require constant attention.[67] The renewed approach remained consistent ever since. For instance, at the 70th anniversary of the KPA's founding on 8 February 2018, Kim Jong-un stressed "cadre army" and "military modernization" as the top priorities in further strengthening the KPA. There is also a great deal of inheritance, where Kim Il-sung is regarded as the architect of the armed forces and the defense planning doctrine, while Kim Jong-il and Kim Jong-un inherited and worked to perfect the developments. Thus, although there were distinctive developments in the KPA since the 1960s, none of them have been inconsistent with the Line of Self-Reliant Defence, and the defense planning doctrine has never receded in its relevance. Rather, all the developments were evolutionary steps that embodied and advanced the Line of Self-Reliant Defence.

The consistency in the DPRK's defense planning is explained by three factors. First, the unchanging nature of the state's grand strategy that continues to focus on the Korean peninsula and the immediate periphery including Japan and the US assets in the western Pacific to achieve unification or at least the survival of the regime. Even though the acquisition of ICBM and SLBM has expanded the KPA's strategic reach, this is different from the DPRK expanding its strategic frontiers to conduct expeditionary operations beyond its region or becoming involved in offshore security matters. Second, the fundamentals of the DPRK's operational concepts and doctrines also remain largely unchanged. Although the strategic arsenals have certainly strengthened the KPA, such assets purport to compensate the KPA's weaknesses in power projection while also adding new dimensions to the DPRK's capabilities for hybrid warfare. Third, the capabilities attained through the Line of Self-Reliant Defence were deemed satisfactory and were the best that the DPRK could achieve within its defense planning framework and also capacity. Hence, there was very little reason for the DPRK to alter its defense planning, as any major reconfigurations could cause various disruptions and even undermine the regime's command and control.

For the DPRK, the Line of Self-Reliant Defence was both a blessing and a curse. On the one hand, the doctrine provided guidelines to strengthen the KPA's readiness while enhancing the regime's centralized and politicized command and control. On the other hand, its far-reaching but path-dependent nature railroaded the military's readiness, creating major shortfalls in force structural and operational readiness. Yet although the Line of Self-Reliant Defence proved to be extremely resource-demanding and cost-ineffective in many aspects, in the eyes of Pyongyang it was essentially the only way to achieve its strategic aims and objectives.

Notes

1 See: Il-sung Kim, "haebangdoen jogukeseoui dang, gukga mit muryeok geonseole daehayeo [On Founding the Party, State and Armed Forces in the Liberated Homeland] (20 August 1945)," in *Kim Il Sung jeojakjib [Kim Il Sung Works]*, vol. 1 (Pyongyang, DPRK: Joseon Rodongdang Chulpansa, 1979), 250–68.

2 See: Il-sung Kim, "hyeonjeongsewa dangmyeongwaeob [The Present Situation and the Immediate Tasks] (21 December 1950)," in *Kim Il Sung jeojakjib [Kim Il Sung Works]*, vol. 6 (Pyongyang, DPRK: Joseon Rodongdang Chulpansa, 1981). The meeting often termed as the "Byeolori meeting" as it was held in Byeolori Village in Jagang Province.

3 See: Baekgwasajeon Chulpansa, *joseon daebaekgwasajeon [Korea Encyclopedia]*, vol. 17 (Pyongyang, DPRK: Baekgwasajeon Chulpansa, 1995), 113, 657, 658; Sahoigwahak Chulpansa, *jeongchisajeon [Dictionary of Politics]* (Pyongyang, DPRK: Sahoigwahak Chulpansa, 1973), 98–99.

4 Il-sung Kim, "juchesasangeui gichireul nopi deulgo sahoijueuigeonseoleul deouk dageuchija [Let Us Step Up Socialist Construction Under the Banner of the Juche Idea] (9 September 1978)," in *Kim Il Sung jeojakjib [Kim Il Sung Works]*, vol. 33 (Pyongyang, DPRK: Joseon Rodongdang Chulpansa, 1987).

5 Jong-un Kim, *songuneui gichireul deo nopi chukyeodeulgo choihuseungrireul hyanghayeo himchage ssawonagaja [Let Us March Forward Dynamically Towards Final*

Victory, Holding Higher the Banner of Songun] (15 April 2012) (Pyongyang, DPRK: Joseon Rodongdang Chulpansa, 2013).

6 See: Baekgwasajeon Chulpansa, *joseon daebaekgwasajeon [Encyclopedia of Korea]*, vol. 17, 658.

7 See: Il-sung Kim, "inmingundaereul jiljeokeuro ganghwahayeo ganbugundaero mandeulja [Let Us Make the People's Army a Cadre Army by Strengthening it Qualitatively] (27 May 1954)," in *Kim Il Sung jeojakjib [Kim Il Sung Works]*, vol. 8 (Pyongyang, DPRK: Joseon Rodongdang Chulpansa, 1980).

8 Baekgwasajeon Chulpansa, *joseon daebaekgwasajeon [Encyclopedia of Korea]*, vol. 3 (Pyongyang, DPRK: Baekgwasajeon Chulpansa, 1995), 271.

9 Kim Il-sung himself gave a number of speeches on the KPA Party Committee system and the "unity" between the commanders and political officers. See: Il-sung Kim, "inmingundaeeui jungdaereul ganghwahaja [Let Us Strengthen the Companies of the People's Army] (11 October 1973)," in *Kim Il Sung jeojakjib [Kim Il Sung Works]*, vol. 28 (Pyongyang, DPRK: Joseon Rodongdang Chulpansa, 1984).

10 See: Joseon Inmingun Chongjeongchiguk, *gunsadaniljereul ganghwahame isseoseo gundaenae rodongdangdanchedeuleui jegwaeob [The Task of the Workers' Party of Korea in the Korean People's Army to Strengthen the Unity of the Armed Forces]* (Pyongyang: DPRK: Joseon Inmingun Chongjeongchiguk, 1951).

11 See: Baekgwasajeon Chulpansa, *joseon daebaekgwasajeon [Encyclopedia of Korea]*, vol. 3, 268.

12 ROK Ministry of Unification, *Understanding North Korea* (Seoul, ROK: ROK Ministry of Unification, 2017), 135.

13 Il-sung Kim, "dang, jeonggwongigwan, inmingundaereul ganghwahamyeo sahoijueuidaegeonseoleul deo jalhayeo hyeokmyeongjeok daesabyeoneul seungrijeokeuro majihaja [Let Us Meet a Revolutionary Upheaval Victoriously by Strengthening the Party, Government Organs, and People's Army and Carrying Out Great Socialist Construction More Efficiently] (17 February 1975)," in *Kim Il Sung jeojakjib [Kim Il Sung Works]*, vol. 30 (Pyongyang, DPRK: Joseon Rodongdang Chulpansa, 1986), 68–71.

14 Joseon Inmingun, "inmingundaeeseoneun dangi jesihan 5daehunryeonbangchimgwa 4daehunryeonwonchikeul teuleojuigo baekdueui hunryeon yeolpungeul sechage ilkyeoya handa [The KPA Must Firmly Grip the Five-point Training Policy and Four-point Training Principles of the Party and Strongly Pursue the Baektu Training Spirit]," *Joseon Inmingun*, January 10, 2010.

15 See: Ibid.

16 See: Jong-il Kim, "songuneui gichireul nopi deulgo inmingundaereul ganghwahaneunde gyesok keun himeul neoheulde daehayeo [On Holding Higher the Banner of Songun and Continuing to Put Power into Strengthening the People's Army] (23, 23 January 2005)," in *Kim Jong Il seonjib [Kim Jong Il Selected Works]*, vol. 22, Expanded ed. (Pyongyang, DPRK: Joseon Rodongdang Chulpansa, 2013), 228–30.

17 See: Ibid., 231.

18 See: Dae-keun Yi, *bukhanguneun woae kudetareul haji ana [Why Don't the Korean People's Army Make a Coup]* (Paju, ROK: Hanul Academy, 2003), 180. Beom-chul Shin and Jin-a Kim, "bukhanguneui cheje [The North Korean Military System]," in *bukhangun sikeurit ripoteu [North Korea Military Secret Report]*, ed. Yong-won Yoo, Beom-chul Shin, and Jin-a Kim (Seoul, ROK: Planet Media, 2013), 57–58.

19 Robert A. Scalapino and Chong-sik Lee, *Communism in Korea* (Berkeley, CA: University of California Press, 1972), 964–65.

20 US Department of the Army, *ATP 7-100.2: North Korean Tactics* (Washington, DC: US Department of the Army, July 2020), 1.13.

21 Ibid., 1.14–1.16.
22 Ibid., 1.16–1.17.
23 See: Il-sung Kim, "jogukhaebangjeonjaengeui jonggukjeokseungrireul irukhagi wihayeo inmingundaeape naseoneun myeotgaji gwaeob [Some Tasks Confronting the People's Army in Winning the Final Victory in the Fatherland Liberation War] (7 February 1952)," in *Kim Il Sung jeojakjib [Kim Il Sung Works]*, vol. 7 (Pyongyang, DPRK: Joseon Rodongdang Chulpansa, 1980); Il-sung Kim, "joseonrodongdang je5cha daehoieseo han jungangwiwonhoisaeobchonghwabogo [Report to the Fifth Congress of the Workers' Party of Korea on the Work of the Central Committee] (2 November 1970)," in *Kim Il Sung jeojakjib [Kim Il Sung Works]*, vol. 25 (Pyongyang, DPRK: Joseon Rodongdang Chulpansa, 1983).
24 Zedong Mao, "On Protracted War," in *Selected Works of Mao Tse-tung*, vol. 2 (Beijing, China: Foreign Languages Press, May 1938), 143.
25 For example, a North Korean encyclopedia describes "military technology" as not only the armaments, infrastructures, and material, but also the personnel's techniques in utilizing those assets. See: Baekgwasajeon Chulpansa, *joseon daebaekgwasajeon [Encyclopedia of Korea]*, vol. 3, 269. Also see: Baekgwasajeon Chulpansa, *joseon daebaekgwasajeon [Encyclopedia of Korea]*, vol. 17, 658.
26 For example, Evans argued that "Modern armies need to find an optimum balance between technology, doctrine and organizational methods at all levels of war. Without adequate doctrine and organization to meet the demands of information systems, warfighting methods risk becoming narrowly delimited by new weapons technology." Michael Evans, *The Continental School of Strategy: The Past, Present and Future of Land Power* (Duntroon, Australia: Land Warfare Studies Centre, 2004), 5.
27 Taik-young Hamm, *Arming the Two Koreas: State, Capital and Military Power* (London, UK and New York: Routledge, 1999), 144. The Korean version of Kim Il-sung's speech found in: Min-ryong Lee, *Kim Jong-il chejeeui bukhangundae haebu [Anatomy of the Kim Jong-il Regime's North Korean Army]* (Seoul, ROK: Hwanggeumal, 2004), 48. Also see: Scalapino and Lee, *Communism in Korea*, 969–73.
28 Il-sung Kim, "sahoijueuigeonseoleui saeroun yogue matge gisulinjaeyangseongsaeobeul ganghwahaja [Let Us Strengthen the Training of Technical Personnel to Meet the New Requirements of Socialist Construction] (2 October 1968)," in *Kim Il Sung jeojakjib [Kim Il Sung Works]*, vol. 23 (Pyongyang, DPRK: Joseon Rodongdang Chulpansa, 1983), 1–2.
29 US Senate Committee on Governmental Affairs Subcommittee on International Security, Proliferation and Federal Services, *Senate Hearing 105–241: North Korean Missile Proliferation*, October 21, 1997.
30 Kwi-geun Kim, "buk, seohae sanggong AN-2giseo misail 2 bal balsa [North Korea – AN-2 Fires Two Missiles in the West Sea]," *Yonhap News*, October 9, 2008, www.yonhapnews.co.kr/bulletin/2008/10/09/0200000000AKR20081009046500043.HTML.
31 See: Baekgwasajeon Chulpansa, *joseon daebaekgwasajeon [Encyclopedia of Korea]*, vol. 3, 271.
32 See: Lee, *Kim Jong-il chejeeui bukhangundae haebu [Anatomy of the Kim Jong-il Regime's North Korean Army]*, 39–40.
33 Hamm, *Arming the Two Koreas: State, Capital and Military Power*, 143.
34 According to Kwon, only those in communications, engineering, mechanized, radar, transport units are eligible to remain in the KPA as non-commissioned officers after their mandatory term. See: Yang-ju Kwon, *bukhangunsaeui ihae [The Comprehension of North Korean Military]*, Expanded ed. (Seoul, ROK: Korea Institute of Defense Analyses, 2014), 139.

35 Ibid., 136–37.
36 ROK Ministry of National Defense, *Defense White Paper 2018* (Seoul, ROK: ROK Ministry of National Defense, 2018), 35.
37 Ibid.
38 Ibid.
39 Scalapino and Lee, *Communism in Korea*, 946.
40 For example, Betts outlined the dichotomy between force structural readiness and operational readiness, where states may face the hard decision of having either a small force capable of quick mobilization, or a large force that may be too slow. Richard K. Betts, *Military Readiness: Concepts, Choices, Consequences* (Washington, DC: Brookings Institution Press, 1995), 45–50.
41 Kim, "joseonrodongdang je5cha daehoieseo han jungangwiwonhoisaeobchonghwabogo [Report to the Fifth Congress of the Workers' Party of Korea on the Work of the Central Committee] (2 November 1970)."
42 Lee, *Kim Jong-il chejeeui bukhangundae haebu [Anatomy of the Kim Jong-il Regime's North Korean Army]*, 42–45; Barbara Demick, "North Korea's Ace in the Hole," *Los Angeles Times*, November 14, 2003.
43 Sung-man Park, "bukhan bidaechingjeonryeoke daehan hangukgun daeeungbangan: jihasiseol muryokhwareul jungshimeuro [The Response of the Korean Armed Forces' Against North Korea's Asymmetric Capabilities: Focus on Neutralizing Underground Facilities]," *habcham [Joint Chiefs of Staff]* 64 (2015): 49.
44 For information on the KPAAF and KPAN's underground bases, see: Joseph S. Bermudez Jr., *The Armed Forces of North Korea* (St. Leonards, Australia: Allen & Unwin, 2001): 100, 138–41.
45 Park, "bukhan bidaechingjeonryeoke daehan hangukgun daeeungbangan: jihasiseol muryokhwareul jungshimeuro [The Response of the Korean Armed Forces' Against North Korea's Asymmetric Capabilities: Focus on Neutralizing Underground Facilities]," 49.
46 Bermudez Jr., *The Armed Forces of North Korea*, 133.
47 CIA National Foreign Assessment Center, *Korea: The Economic Race Between the North and the South* (Washington, DC: CIA, National Foreign Assessment Center, 1978), 6.
48 Joseph S. Bermudez Jr., "MiG-29 in KPAF Service," *KPA Journal* 2, no. 4 (April 2011), 6.
49 Ji You, *China's Military Transformation: Politics and War Preparation* (Cambridge, UK and Malden, MA: Polity Press, 2016), 3.
50 Kwon, *bukhangunsaeui ihae [The Comprehension of North Korean Military]*, 165–66.
51 International Institute for Strategic Studies, *The Military Balance 1961* (London, UK: International Institute for Strategic Studies, 1961), 7.
52 Even when relations with the USSR sourced, some contact was maintained, and Moscow continued to give Pyongyang modest support for the missile program. See: Bermudez Jr., *The Armed Forces of North Korea*, 115–16.
53 Lee, *Kim Jong-il chejeeui bukhangundae haebu [Anatomy of the Kim Jong-il Regime's North Korean Army]*, 330.
54 International Institute for Strategic Studies, *The Military Balance 1971* (London, UK: International Institute for Strategic Studies, 1971), 47–48.
55 See: Kim, "joseonrodongdang je5cha daehoieseo han jungangwiwonhoisaeobchonghwabogo [Report to the Fifth Congress of the Workers' Party of Korea on the Work of the Central Committee] (2 November 1970)," 256.
56 Bermudez Jr., *The Armed Forces of North Korea*, 157; Sung-bin Choi, Jae-moon Yoo, and Si-woo Kwak, *bukhan gunsusaneob gaehwang [Current State of the North Korean Military Industry]* (Seoul, ROK: Korea Institute for Defense Analyses,

2005), 31; Shin and Kim, "bukhanguneui unyeonggwa gunsusaneob [North Korea's Military Management and the Military Industry]," 369.

57 Joseon Jungang Tongshinsa, *joseon jungang nyeongam 1979 [Korea Central Yearbook 1979]* (Pyongyang, DPRK: Joseon Jungang Tongshinsa, 1979), 201.

58 Bermudez Jr., *The Armed Forces of North Korea*, 147.

59 Joseph S. Bermudez Jr., *A History of Ballistic Missile Development in the DPRK*, Occasional Paper (Monterey, CA: Monterey Institute of International Studies Center for Nonproliferation Studies, 1999), 10–11.

60 International Institute for Strategic Studies, *The Military Balance 1981* (London, UK: International Institute for Strategic Studies, 1981), 82.

61 ROK Military Academy, *bukhanhak [North Korea Studies]* (Seoul, ROK: Hwanggeumal, 2006), 226.

62 International Institute for Strategic Studies, *The Military Balance 1991* (London, UK: International Institute for Strategic Studies, 1991), 167.

63 Bermudez Jr., *The Armed Forces of North Korea*, 116.

64 For instance, speaking to the US Senate Select Committee on Intelligence on 11 February 2003, the US Defense Intelligence Agency stated that the Taepodong-1 was a "test bed for multi-stage missile capabilities." US Senate Select Committee on Intelligence, *World Wide Threat Hearing*, February 11, 2003, 7.

65 Kwon, *bukhangunsaeui ihae [The Comprehension of North Korean Military]*, 171–72.

66 Joseon Inmingun, "joseoninmingun choigosaryeonggwan Kim Jong Il dongjiggeseo joseoninmingun ryukgonghaehapdonghunryeoneul bosiyeotda [KPA Supreme Commander Comrade Kim Jong Il Watched Combined Maneuvers of the KPA Three Services]," *Joseon Inmingun*, January 18, 2010.

67 Lee, *Kim Jong-il chejeeui bukhangundae haebu [Anatomy of the Kim Jong-il Regime's North Korean Army]*, 39.

References

Baekgwasajeon Chulpansa. *joseon daebaekgwasajeon [Korea Encyclopedia]*. Vol. 3. Pyongyang, DPRK: Baekgwasajeon Chulpansa, 1995a.

———. *joseon daebaekgwasajeon [Korea Encyclopedia]*. Vol. 17. Pyongyang, DPRK: Baekgwasajeon Chulpansa, 1995b.

Bermudez Jr., Joseph S. *A History of Ballistic Missile Development in the DPRK*. Occasional Paper. Monterey, CA: Monterey Institute of International Studies Center for Nonproliferation Studies, 1999.

———. *The Armed Forces of North Korea*. St. Leonards, Australia: Allen & Unwin, 2001.

———. "SIGINT, EW, and EIW in the Korean People's Army: An Overview of Development and Organization." In *Bytes and Bullets in Korea*, edited by Alexandre Y. Mansourov. Honolulu, HI: APCSS, 2005.

———. "MiG-29 in KPAF Service." *KPA Journal* 2, no. 4 (April 2011).

———. "North Korea Drones On: Redeux." *38 North*, January 19, 2016. www.38north.org/2016/01/jbermudez011916/.

Betts, Richard K. *Military Readiness: Concepts, Choices, Consequences*. Washington, DC: Brookings Institution Press, 1995.

Choi, Sung-bin, Jae-moon Yoo, and Si-woo Kwak. *bukhan gunsusaneob gaehwang [Current State of the North Korean Military Industry]*. Seoul, ROK: Korea Institute for Defense Analyses, 2005.

CIA National Foreign Assessment Center. *Korea: The Economic Race Between the North and the South.* Washington, DC: CIA, National Foreign Assessment Center, 1978.

Demick, Barbara. “North Korea’s Ace in the Hole.” *Los Angeles Times*, November 14, 2003.

Evans, Michael. *The Continental School of Strategy: The Past, Present and Future of Land Power.* Duntroon, Australia: Land Warfare Studies Centre, 2004.

Hamm, Taik-young. *Arming the Two Koreas: State, Capital and Military Power.* London, UK and New York, NY: Routledge, 1999.

International Institute for Strategic Studies. *The Military Balance 1961.* London, UK: International Institute for Strategic Studies, 1961.

———. *The Military Balance 1971.* London, UK: International Institute for Strategic Studies, 1971.

———. *The Military Balance 1981.* London, UK: International Institute for Strategic Studies, 1981.

———. *The Military Balance 1991.* London, UK: International Institute for Strategic Studies, 1991.

Joseon Inmingun. “inmingundaeeseoneun dangi jesihan 5daehunryeonbangchimgwa 4daehunryeonwonchikeul teuleojuigo baekdueui hunryeon yeolpungeul sechage ilkyeoya handa [The KPA Must Firmly Grip the Five-point Training Policy and Four-point Training Principles of the Party and Strongly Pursue the Baektu Training Spirit].” *Joseon Inmingun*, January 10, 2010a.

———. “joseoninmingun choigosaryeonggwan Kim Jong Il dongjiggeseo joseoninmingun ryukgonghaehapdonghunryeoneul bosiyeotda [KPA Supreme Commander Comrade Kim Jong Il Watched Combined Maneuvers of the KPA Three Services].” *Joseon Inmingun*, January 18, 2010b.

Joseon Inmingun Chongjeongchiguk. *gunsadaniljeleul ganghwahame isseoseo gundaenae rodongdangdanchedeureui jegwaeob [The Task of the Workers’ Party of Korea in the Korean People’s Army to Strengthen the Unity of the Armed Forces].* Pyongyang: DPRK: Joseon Inmingun Chongjeongchiguk, 1951.

Joseon Jungang Tongshinsa. *joseon jungang nyeongam 1979 [Korea Central Yearbook 1979].* Pyongyang, DPRK: Joseon Jungang Tongshinsa, 1979.

Kim, Il-sung. “haebangdoen jogukeseoui dang, gukga mit muryeok geonseole daehayeo [On Founding the Party, State and Armed Forces in the Liberated Homeland] (20 August 1945).” In *Kim Il Sung jeojakjib [Kim Il Sung Works].* Vol. 1. Pyongyang, DPRK: Joseon Rodongdang Chulpansa, 1979.

———. “inmingundaereul jiljeokeuro ganghwahayeo ganbugundaero mandeulja [Let Us Make the People’s Army a Cadre Army by Strengthening it Qualitatively] (27 May 1954).” In *Kim Il Sung jeojakjib [Kim Il Sung Works].* Vol. 8. Pyongyang, DPRK: Joseon Rodongdang Chulpansa, 1980a.

———. “jogukhaebangjeonjaengeui jonggukjeokseungrireul irukhagi wihayeo inmingundaeape naseoneun myeotgaji gwaeob [Some Tasks Confronting the People’s Army in Winning the Final Victory in the Fatherland Liberation War] (7 February 1952).” In *Kim Il Sung jeojakjib [Kim Il Sung Works].* Vol. 7. Pyongyang, DPRK: Joseon Rodongdang Chulpansa, 1980b.

. “hyeonjeongsewa dangmyeongwaeob [The Present Situation and the Immediate Tasks] (21 December 1950).” In *Kim Il Sung jeojakjib [Kim Il Sung Works].* Vol. 6. Pyongyang, DPRK: Joseon Rodongdang Chulpansa, 1981.

———. "joseonrodongdang je5cha daehoieseo han jungangwiwonhoisaeobchonghwabogo [Report to the Fifth Congress of the Workers' Party of Korea on the Work of the Central Committee] (2 November 1970)." In *Kim Il Sung jeojakjib [Kim Il Sung Works]*. Vol. 25. Pyongyang, DPRK: Joseon Rodongdang Chulpansa, 1983a.

———. "sahoijueuigeonseoleui saeroun yogue matge gisulinjaeyangseongsaeobeul ganghwahaja [Let Us Strengthen the Training of Technical Personnel to Meet the New Requirements of Socialist Construction] (2 October 1968)." In *Kim Il Sung jeojakjib [Kim Il Sung Works]*. Vol. 23. Pyongyang, DPRK: Joseon Rodongdang Chulpansa, 1983b.

———. "inmingundaeeui jungdaereul ganghwahaja [Let Us Strengthen the Companies of the People's Army] (11 October 1973)." In *Kim Il Sung jeojakjib [Kim Il Sung Works]*. Vol. 28. Pyongyang, DPRK: Joseon Rodongdang Chulpansa, 1984.

———. "dang, jeonggwongigwan, inmingundaereul ganghwahamyeo sahoijueuidaegeonseoleul deo jalhayeo hyeokmyeongjeok daesabyeoneul seungrijeokeuro majihaja [Let Us Meet a Revolutionary Upheaval Victoriously by Strengthening the Party, Government Organs, and People's Army and Carrying Out Great Socialist Construction More Efficiently] (17 February 1975)." In *Kim Il Sung jeojakjib [Kim Il Sung Works]*. Vol. 30. Pyongyang, DPRK: Joseon Rodongdang Chulpansa, 1986.

———. "juchesasangeui gichireul nopi deulgo sahoijueuigeonseoleul deouk dageuchija [Let Us Step Up Socialist Construction Under the Banner of the Juche Idea] (9 September 1978)." In *Kim Il Sung jeojakjib [Kim Il Sung Works]*. Vol. 33. Pyongyang, DPRK: Joseon Rodongdang Chulpansa, 1987.

Kim, Jong-il. "songuneui gichireul nopi deulgo inmingundaereul ganghwahaneunde gyesok keun himeul neoheulde daehayeo [On Holding Higher the Banner of Songun and Continuing to Put Power into Strengthening the People's Army] (23, 23 January 2005)." In *Kim Jong Il seonjib [Kim Jong Il Selected Works]*. Vol. 22. Expanded ed. Pyongyang, DPRK: Joseon Rodongdang Chulpansa, 2013.

Kim, Jong-un. *songuneui gichireul deo nopi chukyeodeulgo choihuseungrireul hyanghayeo himchage ssawonagaja [Let Us March Forward Dynamically Towards Final Victory, Holding Higher the Banner of Songun] (15 April 2012)*. Pyongyang, DPRK: Joseon Rodongdang Chulpansa, 2013.

Kim, Kwi-geun. "buk, seohae sanggong AN-2giseo misail 2 bal balsa [North Korea – AN-2 Fires Two Missiles in the West Sea]." *Yonhap News*, October 9, 2008. www.yonhapnews.co.kr/bulletin/2008/10/09/0200000000AKR20081009046500043.HTML.

Kwon, Yang-ju. *bukhangunsaeui ihae [The Comprehension of North Korean Military]*. Expanded ed. Seoul, ROK: Korea Institute of Defense Analyses, 2014.

Lee, Min-ryong. *Kim Jong-il chejeeui bukhangundae haebu [Anatomy of the Kim Jong-il Regime's North Korean Army]*. Seoul, ROK: Hwanggeumal, 2004.

Mao, Zedong. "On Protracted War." In *Selected Works of Mao Tse-tung*. Vol. 2. Beijing, China: Foreign Languages Press, May 1938.

Park, Sung-man. "bukhan bidaechingjeonryeoke daehan hangukgun daeeungbangan: jihasiseol muryokhwareul jungshimeuro [The Response of the Korean Armed Forces' against North Korea's Asymmetric Capabilities: Focus on Neutralizing Underground Facilities]." *habcham [Joint Chiefs of Staff]* 64 (2015).

ROK Military Academy. *bukhanhak [North Korea Studies]*. Seoul, ROK: Hwanggeumal, 2006.

ROK Ministry of National Defense. *Defense White Paper 2018*. Seoul, ROK: ROK Ministry of National Defense, 2018.

ROK Ministry of Unification. *Understanding North Korea*. Seoul, ROK: ROK Ministry of Unification, 2017.

Sahoigwahak Chulpansa. *jeongchisajeon [Dictionary of Politics]*. Pyongyang, DPRK: Sahoigwahak Chulpansa, 1973.

Scalapino, Robert A., and Chong-sik Lee. *Communism in Korea*. Berkeley, CA: University of California Press, 1972.

Shin, Beom-chul, and Jin-a Kim. "bukhanguneui cheje [The North Korean Military System]." In *bukhangun sikeurit ripoteu [North Korea Military Secret Report]*, edited by Yong-won Yoo, Beom-chul Shin and Jin-a Kim. Seoul, ROK: Planet Media, 2013a.

———. "bukhanguneui unyeonggwa gunsusaneob [North Korea's Military Management and the Military Industry]." In *bukhangun sikeurit ripoteu [North Korea Military Secret Report]*, edited by Yong-won Yoo, Beom-chul Shin, and Jin-a Kim. Seoul, ROK: Planet Media, 2013b.

US Department of the Army. *ATP 7–100.2: North Korean Tactics*. Washington, DC: US Department of the Army, July 2020.

US Senate Committee on Governmental Affairs Subcommittee on International Security, Proliferation and Federal Services. *Senate Hearing 105–241: North Korean Missile Proliferation*, October 21, 1997.

US Senate Select Committee on Intelligence. *World Wide Threat Hearing*, February 11, 2003.

Yi, Dae-keun. *bukhanguneun woae kudetareul haji ana [Why Don't the Korean People's Army Make a Coup]*. Paju, ROK: Hanul Academy, 2003.

You, Ji. *China's Military Transformation: Politics and War Preparation*. Cambridge, UK and Malden, MA: Polity Press, 2016.

6 The state of readiness

For the military to effectively execute their tasks, states need to ensure the optimum balance between force structural and operational readiness. While personnel and hardware such as aircraft, vehicles, vessels, and various equipment and instruments shape the structural capabilities, one must also consider the state of supplies, maintenance, personnel, education, and training that determine how the capabilities will be mobilized and operated. Moreover, military readiness today is not simply about fielding the best equipment and weapons, but about how they are coordinated and integrated through systems such as C4ISTAR. Combined with the increasing sophistication in military operations, readiness management has become more complex, requiring states to make sensible decisions in ensuring effectiveness and efficiency. Still, as discussed in Chapter 5, efforts to attain and sustain that balance are often hamstrung by over-politicized interests, limited economic and industrial capacity, as well as simple poor management. Consequently, states often end up with not only compromised level of military readiness that creates challenges in achieving their strategic aims and objectives but also narrows opportunities for reconfiguration and improvement.

Although all states face problems in attaining and sustaining optimal readiness, the problems in the DPRK are especially dire. Indeed, Pyongyang found a formula that focuses on the military's integrity and loyalty while enhancing the KPA's readiness. One could argue that the DPRK was wise in avoiding the ambitious acquisition of heavy-duty, state-of-the-art platforms that would implicate high investment and O&M costs that are beyond the state's capacity. Still, the way in which the DPRK handled its defense planning dilemmas led to major compromises in the military's force structural and operational readiness. Moreover, the troubled economic circumstances left serious long-term ramifications on the operational readiness of the KPA. Thus, despite Pyongyang's efforts to strengthen the KPA in accordance with the Line of Self-Reliant Defence, the gains were outweighed by the myriad problems that not only undermined their ability to effectively conduct defensive and offensive operations but also in fighting long conflicts.

Force structural readiness

The KPA's force structure accurately reflects the state's defense planning that focuses on hybrid warfare. Over time, Pyongyang focused on a quantitative force structure for asymmetric operations rather than qualitative pursuit for high-end technologies. While the approach allowed the DPRK to build a massed force, it failed to achieve a balanced and coherent modernization of the KPA. A significant portion of the KPA inventory is antiquated by today's standards, and even the most modern platforms are at least a generation or two behind those of neighboring states. As time passed, the portion of outdated platforms in the KPA inventory grew, revealing how the DPRK has failed to adequately modernize the armed forces. Indeed, the DPRK is not leaving the KPA inventory to rot, epitomized by its WMD program as well as some incremental modernization of conventional weaponry. Even with the hardware that is barely operational, many are used as decoys or live shields to impose costs on the enemy. Nevertheless, there are notable problems in the KPA inventory, particularly in conventional capabilities that undermine the military's overall readiness.

Ground warfare

Since its inception, ground warfare has been the backbone of the KPA. Currently, the ground component of the KPA is known to have approximately 1,100,000 active personnel with ten regular infantry corps, two mechanized corps, Pyongyang Defence Command, and the special operations commands.[1] The bulk of the KPA's ground forces are forward-deployed, fielding an array of infantry, artillery, MLRS, and armored units below the Pyongyang–Wonsan line for offensive raids into South Korea as well as acting as a tripwire against incoming forces.

One of the most threatening assets of the KPA is the 21,600-plus artillery and MLRS units capable of firing both conventional and chemical-tipped munitions.[2] At present, the KPA's self-propelled artillery units include: 122mm M-1977, M-1981, M-1985, and M-1991; 130mm M-1975, M-1981, and M-1991; 152mm M-1974, M-1977, and the M-2018; as well as the 170mm M-1978 and M-1989.[3] Towed artillery includes 122mm D-30, D-74, M-1931/37; 130mm M-46; and 152mm M-1937, M-1938, M-1943.[4] As for MLRS, much of the inventory consists of the 107mm Type-63; 122mm BM-11, M-1977 (BM-21), M-1985, M-1992, and M-1993; 200mm BMD-20; 240mm BM-24, M-1985, M-1989, and M-1991; and the 300mm KN-09.[5] The KN-09 is the most modern and powerful, resembling the Chinese A-100 or Russian BM-30, and boasts a range of approximately 200 km.[6] Combined with the range of mortars, the KPA's artillery and MLRS units pose significant threats as many of them are forward-deployed literally holding ROK and US assets near the border (including the Seoul metropolitan area) at gunpoint and are also garrisoned in HARTS for surprise attacks while also being protected from enemy fire.

The KPA is also known to possess over 4,100 main battle tanks and light tanks.[7] Much of the KPA's main battle tanks are dated imports from the USSR and China such as the T-34, T-54, T-55, T-62, Type-59, and Type-62 (Chinese variants of the T-54 and T-59, respectively) as well as amphibious light tanks such as the PT-76 and Type-63. Over the last several decades, many of the vintage tanks have been replaced by the more modern platforms. From the 1970s, the DPRK embarked on the indigenous production of main battle tanks and light tanks based on Chinese and Soviet imports. The Cheonma-ho unveiled in the 1980s is based on the Soviet 115mm-gunned T-62, and the PT-85 Shinheung amphibious light tanks has some resemblance with the PT-76. In the early 1990s, the KPA began operating the Pokpung-ho that is essentially an upgraded version of the Cheonma-ho fitted with fire control systems and laser rangefinders.[8] Sometime in the 2000s, the DPRK unveiled the Songun-ho that is based on the T-62 but fitted with a range of modern equipment including the 125mm tank gun, 14.5mm machine gun, anti-tank missile launchers, man-portable air defense systems, upgraded fire control systems, infrared sensors, and laser rangefinders.[9] The most recent main battle tank was introduced in October 2020, with physical features that have features resembling not only the Russian T-14 Armata but also to some extent the Chinese VT-4 and even the US M1 Abrams. Even though much of the armored combat vehicles of the KPA are dated, they are quantitatively sufficient in causing significant damage against the ROK and US forces in South Korea in the initial stages of battle.

The KPA fields an array of armored personnel carriers and armored infantry fighting vehicles. A large number of vehicles include those imported from China and the USSR during the Cold War, including the BTR-40, BTR-50, BTR-60, BTR-152, and Type-63.[10] Yet as the DPRK's vehicle production capacity strengthened, the imports have been phased out by domestic variants including the M-1992 based on the Soviet BRDM-2 and the VTT-323 based on the Chinese Type-63 capable of functioning as an amphibious assault vehicle.[11] The newest model is the M-2010 fitted with fire control systems that comes in both six- and eight-axel variants that seem to be based more on the BTR-60 and BTR-80.[12] In addition, the DPRK also showcased a new, compact 4WD armored vehicle at a military parade in September 2018 that could be fitted with light arms. The armored vehicles allow the KPA with greater flexibility and maneuverability for transportation and fire support albeit questions over the durability of the armor against modern anti-tank weaponry.

The KPA has also placed much focus on anti-tank missiles. The KPA has a number of anti-tank ordnances such as the 9M14 (AT-3 Sagger) as well as man-portable anti-tank systems such as the 2K15 (AT-1 Snapper), 9K111 (AT-4 Spigot), and the 9K113 (AT-5 Spandrel).[13] In recent years, the DPRK also produced its own laser-guided anti-tank missiles known as the Bulsae series based on the 9K111 (AT-4 Spigot) and a newer model based on the 9M133 (AT-14 Spriggan).[14] In addition, the KPA also has an assortment of anti-personnel and anti-vehicle mines, booby-traps, and improvised explosive devices to be used against enemy ground units.

Another powerful asset of the KPA is the special operations units specialized in commando raids and operations behind enemy lines, including assassinations, abductions, reconnaissance, sabotage of critical civilian and military infrastructures, sniping, counterstability operations, as well as bodyguarding the leader.[15] Estimates on the total number of special operations personnel in the KPA vary from 88,000 to 200,000.[16] The statistical gap may come from confusions over the number of those in the special operations units versus soldiers trained for special operations. Moreover, it is possible that high-end estimates simply point to the number of combat-ready infantry soldiers. It is generally understood that the special operations units are dispersed across the KPA, with most affiliated to the 11th Corps (formerly known as the Light Infantry Training Guidance Bureau), the RGB, the Guard Command, as well as the KPAN and the KPAAF for amphibious and airborne operations. Yet in April 2017, the DPRK reportedly established a new command known as the KPASOF headed by Colonel General Kim Yong-bok.[17] It is unknown, however, whether the KPASOF is simply an enlarged and renamed 11th Corps, or whether it is a command that brings together the various special operations units. Upgrades in equipment are also evident. At the military parade on 15 April 2017 to mark the "Day of the Sun," the DPRK revealed a unit of soldiers wearing fatigues and body armor with digital camouflage patterns similar to those worn by the ROK Army, as well as night-vision goggles and Type-98 assault rifles fitted with helical magazines.

The KPA has a number of helicopters for close air support, reconnaissance, and surveillance. Yet the only attack helicopter in the KPA is the Mi-24 helicopter gunship capable of light troop transportation.[18] Otherwise, much of the DPRK's helicopters are transport and utility platforms fitted with armaments. For instance, the Mi-2 is armed with machine guns and anti-tank rockets.[19] Another example is the 80-odd MD-500E fitted with various armaments and painted with markings of the ROK forces to create confusion while engaging in combat, reconnaissance, and transportation of commandos.[20] The relatively small number of attack helicopters speaks to the fuel shortages in the DPRK. Nevertheless, the various helicopters would still play vital roles in close air support, reconnaissance, and transportation for the KPA's ground operations.

Despite using dated inventory, it is important to note that ground warfare is much less dependent on technology and less cost-intensive than in other domains. As witnessed in various wars around the world, troops armed with dated equipment but highly fluent in asymmetric tactics have been effective against technologically superior opponents. The building of a massed army designed for barrage, swarm, asymmetric attacks at moment's notice compensates for the KPA's technological deficiencies against the ROK and US counterparts. Such features have been the reason why readiness for ground warfare has been the mainstay of the KPA – both for offense and defense. Still, the KPA's heavy focus on ground warfare creates shortfalls, particularly as both the US and ROK have grown familiar with the threats and also enhanced their countermeasure and strike capabilities. Hence, although there is no doubt about the KPA's ability to inflict significant damage and disruption to its southern brethren, its actual effectiveness to dictate the outcome of conflict is questionable.

Naval warfare

The KPAN operates two fleets with the East Sea Fleet headquartered in Rakwon and the West Sea Fleet headquartered in Nampho. While the KPAN only has approximately 60,000 personnel, it boasts a high number of light surface vessels and submarines built for offensive and defensive operations in the littorals.[21] Due to the combination of resource and technological constraints, much of the KPAN inventory is obsolete or technologically behind modern standards. Still, recent developments show that the DPRK is making some efforts to modernize their naval capabilities, particularly with the construction of new submarines, as well as some new and converted surface vessels capable of launching the Kumsong-3 ASCM based on the Kh-35U (AS-20 Kayak).[22]

The major strength of the KPAN is the large fleet of submarines. The largest attack submarine is the 20-odd 1,830t Romeo-class that has been in service since the 1970s.[23] Numerically, much of the KPAN submarine fleet consists of compact submarines built for asymmetric attacks, reconnaissance, and the infiltration/exfiltration of commandos. The core of the attack submarine fleet is the 370t Sango-I produced in the 1990s followed by the larger Sango-II with two additional torpedo tubes developed in the 2000s.[24] Moreover, the KPAN has invested in a number of midget submarines starting with the 90t Yugo-class first acquired in the 1960s and the 130t Yeono-class acquired in the 1990s.[25] The small size of the KPAN submarines has often served as an advantage for the DPRK epitomized by the numerous infiltrations into South Korea and surprise attacks like the sinking of the ROK Navy corvette Cheonan in March 2010.

Mysteries remain regarding the developments in ballistic missile submarines (SSB) and/or nuclear submarines (SSBN). In 2014, reports claimed that the DPRK launched the conventionally powered Sinpo-class (also known as Gorae-class) capable of launching only one SLBM. Yet the specifications of the Sinpo-class suggested that it is more of an experimental platform, and there were growing reports in 2016 and 2017 that the DPRK is, in fact, working on a new class of SSB referred to as Sinpo-C.[26] Much was confirmed in July 2019 when the North Korean state media propagated Kim Jong-un's visit the Sinpho Naval Shipyard, revealing photographs of the new SSB based on the Romeo-class but fitted with a vertical launching system capable of launching multiple SLBM.[27] Progress seems to be taking place – at least according to Kim Jong-un – claiming at the Eighth WPK Congress in January 2021 that the R&D of a SSBN is well underway. While much remains uncertain, SSB and SSBN will undoubtedly be one of the key agendas for the DPRK's military modernization program that would significantly enhance their strategic capabilities.[28]

Surface warfare has long been the weakness of the KPAN. Much of the KPAN's surface combatants are corvettes, frigates, patrol vessels, and torpedo boats for asymmetric attacks, fire support for amphibious operations and some minelaying operations in littoral areas. Yet there are some questions about the KPAN surface warfare fleet. First, there is much uncertainty about the operational status of the majority of the KPAN's surface combatants given that many are extremely dated imports from the Cold War era or retrofitted variants of those models. Second, identifying KPAN vessels is made difficult by the mix-and-match of armaments and other specifications, often leading to confusions over the categorization of the vessels. Some general trends are evident with newer or retrofitted vessels equipped with the AK-230 30mm close-in weapon system (CIWS), RBU-1200 anti-submarine rocket launchers, and the Kumsong-3 ASCM.

The largest of the KPAN's surface warships include the 1,500t Rajin-class, 1,300t Nampo-class helicopter frigates, and what seems to be a 3,000t Krivak-class frigate.[29] In the mid-2010s, a number of images of what could be the Nampo (or improved variants) revealed a superstructure with reduced radar cross-section with a deck for one helicopter and armed with AK-230 30mm CIWS, Kumsong-3 ASCM, and RBU-1200 anti-submarine rocket launchers.[30] As for corvettes, the KPAN is reported to still operate two pre-1945 vessels such as the 650t Sariwon-class and 580t Tral-class that are lightly armed but fitted with either Pot Head or Don-2 radars and also the Stag Horn sonar system.[31] Other corvettes include the Hainan- and Daecheong-I/II-class equipped with RBU-1200 anti-submarine rocket launchers.[32] The remainder of the KPAN – about 97% of the surface vessels – comprises what could be classed as patrol vessels. Numerically, much of them are fast patrol boats that can be categorized based on armaments. Fast patrol boats armed with guided missiles include the Komar-, Osa-I- (including the Huangfeng-class Chinese variant and the Soju-class DPRK variant), Sohung-, and Nongo-classes while those equipped with simple armaments include the Cheongjin-, Guseong-, Sinheung-, and the Sinpo-classes.[33] The KPAN also possesses a number of the severely dated patrol boats of questionable operational status including the Chaho-, Cheonju, Shanghai-II-, and SO-I-classes.[34]

While the far majority of the KPAN's surface combatants remains to be dated, there has been notable efforts to modernize the inventory with emphasis on vessels with smaller radar cross-sections. One interesting development is the Haesam- and Nongo-class surface effect ships (SES) fitted with capabilities to launch the Kumsong-3 ASCM.[35] While both the Haesam and Nongo have reduced radar cross-sections features, the various armaments and radars fitted atop of the ship compromise the vessels' stealth capabilities. In addition, the DPRK has also commissioned the high-speed very slender vessels (VSV) designed for reconnaissance, transportation, and asymmetric attacks.[36] Indeed, the SES and VSV are light and, therefore, do not indicate the KPAN's shift toward blue-water capabilities. Still, the new assets conform to the KPAN's emphasis on dealing with threats approaching its waters, as well as asymmetric attacks, amphibious, and reconnaissance operations.

The KPAN possesses a wealth of vessels for amphibious and reconnaissance operations. For amphibious assaults, the KPAN possesses approximately 10 Hantae-class landing ships and approximately 257 landing crafts of various types including the Gongban-I/II/III and Nampo-classes.[37] Many of the KPAN's amphibious assets are forward-deployed to facilitate landing of infantry units as well as the infiltration/exfiltration of agents and commandos into the South, with several hovercraft bases on the western coast of the Korean peninsula near the NLL.[38] For covert infiltration and exfiltration as well as reconnaissance, the DPRK also has not only operated high-speed semi-submersible vessels and midget submarines but also retrofitted fishing boats as deceptive and less detectable means to infiltrate South Korean and Japanese waters.[39] Retrofitted civilian vessels have entered Japanese waters on a number of occasions with notable cases in March 1999, December 2001, and September 2002.

Minewarfare is also a critical operation of KPAN. In defense, naval mines would slow and deny enemy access into North Korean waters and ports, while offensive operations involve minelaying in SLOCs and ports in South Korea and Japan. According to Bermudez, the KPAN utilizes the 24-odd Yukto-I/II minesweeping vessels as well as various surface combatants, submarines, and civilian vessels for minelaying.[40] Although there are questions over the effectiveness of civilian vessels, the quantity and variety of the vessels mobilized for minelaying operations are sufficient in creating headaches for opposing forces, forcing them into intensive minesweeping operations before making any advances.

Despite boasting a large inventory, the KPAN has suffered humiliating defeats against the ROK Navy in the naval battles off Yeonpyeong Island in June 1999 and June 2002 as well as Daecheong in November 2009. Much of the problems were due to the armaments, radars, and sonars with limited firepower, precision, and range. Despite some levels of modernization with the acquisition of new platforms in recent years, the far majority of the KPAN inventory are armed with Cold War-grade anti-ship and anti-submarine weaponry such as the 53mm torpedoes or the Soviet P-15/P-20 (SS-N-2 Styx) (including the CSS-C-2 Chinese variant and the KN-01 DPRK variant) and RBU-1200, or simply with cannons and machine guns that make them only suitable for close combat.[41] Indeed, the DPRK has sought to indigenously develop more modern naval weaponry such as anti-ship missiles, torpedoes, and CIWS. Still, much of them are based on older generation models from China and the USSR, making them questionable in their effect against the more advanced systems of the US, ROK, and Japanese counterparts.[42]

Weaknesses are also evident in anti-submarine warfare (ASW), making the KPAN vulnerable against enemy submarines. Patrol boats such as the SO-I, Cheongju, Daecheong-I/II, and Hainan-class are equipped with dated Stag Horn and Stag Ear sonar systems and RBU-1200 anti-submarine rocket launchers that have questionable ASW capabilities by modern standards.[43] Some of the larger vessels such as the Nampo-class are fitted with helicopter decks to accommodate the Mi-14PL acquired sometime in the 1980s and 1990s.[44] However, such vessels are only capable of accommodating one helicopter each, forcing the rest of the helicopters to be deployed from onshore bases. Moreover, the DPRK is not known to possess purpose-built fixed-wing maritime patrol aircraft. Thus, the KPAN's ASW capabilities are qualitatively and quantitatively insufficient, imposing disadvantages to its surface and coastal combatants to maneuver in the seas surrounding the Korean peninsula.

The KPAN's coastal defense systems are built much like the ground forces. The DPRK has built shore-based anti-access systems equipped with coastal artillery and anti-ship missiles of various types including the Kumsong-3.[45] The coastal defense systems are built to protect the naval bases and factories, as well as dealing with enemy amphibious landings. The Kumsong-3 has certainly added extra clout, allowing the KPAN with the shore-based option of engaging enemy fleets (particularly aircraft carriers) in ways that are much less affected by sea and weather conditions. Moreover, the small-scale nature of the DPRK's naval platforms allows them to be bunkered and kept ready in fortified naval bases and defended by the coastal defense systems.

Despite its quantitative inventory, the KPAN has failed to modernize, consequently limiting its capabilities for asymmetric attacks, fire support, transportation, minelaying, and launch of ASCM and SLBM in coastal waters as well as disruption to SLOCs and ports. There is, however, a growing demand for new capabilities, particularly in fast, stealthier vessels, and submarines with the ability to launch SLBMs. In recent years, the DPRK is known to be undertaking a naval modernization program, evidenced by the increased activities in key bases such as Munchon, Nampho, and Sinpho.[46] Going forward, although the KPAN is still far from building blue-water naval capabilities, one could expect the phasing out of old platforms with upgrades to existing operational platforms and production of various technological demonstrators and prototypes of surface vessels, submarines, or even unmanned surface vehicles (USV) and unmanned underwater vehicles (UUV) in the coming years. How far the DPRK can go in developing new platforms, and how they are operated remains uncertain. While the DPRK's track record and its production capacity make it unlikely for the KPAN to become a modern, blue-water navy in the near-term, the trajectory of the current developments nevertheless warrant greater attention.

Air warfare

The KPAAF has approximately 110,000 personnel with four combat divisions based in Hwangju, Kaechon, Orang, and Toksan, as well as two transport divisions based in Sondok and Taechon.[47] The DPRK's quest to build resilient air defense capabilities was based on the lessons from the Korean War where the KPA was overwhelmed by the airpower of the US and its allies. However, although the DPRK sought to enhance its airpower capabilities with much assistance from China and the USSR, it also became the branch that was the most affected by the economic troubles and lacking access to technology, evidenced by the technological deficiencies in the KPAAF inventory.

Currently, the KPAAF is known to operate over 400 tactical fighters. Numerically, the KPAAF's tactical fighter inventory mostly consists of first- and second-generation Soviet platforms and their Chinese variants such as the MiG-15, MiG-17 (or J-5), MiG-19 (or J-6), and MiG-21F-13/PFM (or J-7).[48] In the mid-1980s, the DPRK acquired its first third-generation fighter, the MiG-23ML/P with approximately 56 in the current KPAAF inventory.[49] To date, the only fourth generation tactical fighter in the KPAAF is the 18-odd MiG-29A/S/UB acquired from the USSR in the latter half of the 1980s.[50] In real terms, the MiG-23 and MiG-29 are the only platforms that the KPAAF can count on in aerial combat, but their quantitative deficiencies against the more technologically advanced tactical aircraft and weapons of the US, ROK, and Japanese counterparts nevertheless highlight the constraints in gaining air superiority and even questionable in securing air parity. Moreover, despite their competence in maneuverability and speed, the KPAAF's MiG-29s are primarily affiliated to the 55th Kumsong Guard Air Regiment for the defense of Pyongyang and have been sparingly dispatched for combat air patrols consequently making the MiG-23s serve as the workhorse of the KPAAF.[51]

The KPAAF also has some air-to-ground attack capabilities with the MiG-21bis, A-5 (export variant of the Q-5), Su-7, and the Su-25/Su-25UBK.[52] Among them, the Su-25, often dubbed as the USSR equivalent of the A-10, is arguably the most lethal, heavily armored and equipped with 30mm GSh-30-2 automatic cannon to strafe its target. Regarding bombers, the only known platform in the KPAAF is the Il-28 (or the H-5 Chinese variant) designed primarily for free-fall bombs but have undergone some modifications to launch the KN-01 anti-ship missiles and for electronic warfare.[53] While the KPAAF attempted to develop capabilities for standoff strikes on high-value targets and fire support, they are nonetheless vulnerable to modern-day air-to-air and surface-to-air missiles that boast greater range and precision. Consequently, such limitations undermine the KPAAF's ability to go deep into the opponent's territory, confining them to tactical missions within or near North Korean territory as opposed to strategic strikes against South Korea or Japan.

The critical weakness of the KPAAF rests in the mostly dated air-to-air and air-to-surface missiles with inferior accuracy and operational radius to the current-day counterparts. In short-range infrared homing missiles, the KPAAF has the R-3 (AA-2 Atoll), R-60 (AA-8 Aphid), and R-73 (AA-11 Archer) acquired from Moscow as well as the PL-5 (based on the R-3) and PL-7 (based on the R.500 Magic) from Beijing.[54] As for semi-active radar homing medium-range missiles, the DPRK has the R-23 (AA-7 Apex), R-24 (upgraded version of the R-23), and the medium-to-long-range R-27R/ER (AA-10 A/C Alamo).[55] Even in attacking ground targets, the KPAAF has a stockpile of not only conventional free-fall bombs but also cluster munitions, chemical and biological agents, napalm, smoke bombs, and some air-launched guided missiles.[56] As for air-to-surface missiles, the KPAAF is known to have the Kh-23 (AS-7 Kerry) and the laser-guided Kh-25 (AS-10 Karen) and Kh-29 (AS-14 Kedge).[57] Despite the variety, the armaments are generally dated and even the more modern ones are only compatible with a fraction of the KPAAF inventory. The lack of accuracy and reach means that the KPAAF aircraft face the risk of being shot down by the longer range air-to-air and surface-to-air missiles of the US, ROK, and Japan. Indeed, the DPRK could try to indigenously produce air-launched ordnances like it has done so with the Bulsae anti-tank missiles. Yet the production of air-to-air and air-to-surface missiles would prove to be far more complex due to technological hurdles.

To compensate for the weaknesses in air-to-air capabilities, the DPRK has placed strong emphasis on anti-air and airborne assault defense systems. Currently, the KPA fields a variety of surface-to-air missiles such as the 9K35 (SA-13 Gopher), S-75 (SA-2 Guideline), S-125 (SA-3 Goa), S-200 (SA-5 Gammon), and the Pongae-5 (KN-06) that closely resembles China's HQ-9 and the Russian S-300 (SA-10 Grumble).[58] As for anti-aircraft cannons and guns, the KPA operates a wide variety of towed pieces such as the 14.5mm ZPU-1/2/4, 23mm ZU-23, 37mm M-1992, 57mm S-60, 85mm M-1939, and the 100mm KS-19.[59] The self-propelled units include the 14.5mm M-1984, 23mm M-1992, 37mm M-1992, and the 57mm M-1985.[60] In addition, the DPRK is also known to operate an unknown but large quantity of man-portable air defense systems including the 9K32 (SA-7 Grail), 9K310 (SA-16 Gimlet), 9K38 (SA-18 Grouse), HN-5/5A (Red Tassel), and the FIM-92A (Stinger).[61]

Like the KPAN, the KPAAF places strong emphasis on transportation. For fixed-winged aircraft, the KPAAF has the An-24, Il-76, and about 200 of the retrofitted An-2.[62] As for rotary-winged aircraft, the KPAAF has approximately 286 helicopters with almost all of them for small and medium lift such as the MD-500E, Mi-2, Mi-4, Mi-8, and Mi-17, and four of the colossal Mi-26 that can carry up to 90 personnel.[63] The KPAAF's airlift capabilities are also augmented by the aircraft of Air Koryo such as the An-148, Il-18, Il-62, Tu-134, Tu-154, and Tu-204. In addition, the KPA is also known to have hot air balloons and sailplanes for less detectable ways of infiltrating troops into South Korea.[64] Although the low-tech platforms are only suited for light airlift, they are utilized for swarm infiltration, where the DPRK aims to get only a fraction of the units over the MDL to fulfill their objectives. While such an approach is risky, they are nevertheless one of the few options available.

One of the unique aspects of the KPAAF is the multirole application of platforms. The most prime example is the An-2. While dated, the biplane's ability for short take-off and landing and fly at low altitudes makes it good means to infiltrate commandos, conduct ISTAR operations as well as certain weapons.[65] Several other aircraft are also retrofitted to play supportive roles. For instance, the selected number of the Il-76 is used for aerial refueling and airborne warning and control system (AWACS) operations, and some of the An-24 is known to be used for airborne early warning (AEW).[66] The multirole utilization of transport aircraft reflects the resource and technological constraints in the DPRK that have forced them to settle for compromised alternatives. Hence, even though the DPRK tried to acquire some capabilities in AEW, AWACS, and aerial refueling, the actual effect on the KPAAF is questionable.

Notable developments are seen in UAS for both reconnaissance and strike. The DPRK's UAS program span over decades, starting with the acquisition of the D-4 NPU from China in sometime in the late 1980s and early 1990s, Tu-143 DR-3 from Syria in 1994, Pchela-1T from Russia in 1997, as well as the Sky-09P from China and MQM-107D via the Middle East in sometime in the 2000s and early 2010s.[67] While the imported platforms were not the best available, they nevertheless served as technological templates for the DPRK's indigenous

UAS. Since 2012, the DPRK has showcased some drones at a number of military parades and some military exercises. On numerous occasions, the DPRK's UAS have even crossed into the South for either actual reconnaissance missions or to test the drones' effectiveness in enemy territory, with some crashing and revealing the simple and rudimentary specifications that were a mix-and-match of commercial and dual-use components.

For the DPRK, UAS are vital assets given their ability to operate in hazardous areas and maneuver in ways beyond human physiological capacity. Even though the KPA's drones are still in the early stages of development, Pyongyang would seek to enhance their durability, firepower, range, stealth, and ISTAR capabilities. Given the problems in the tactical aircraft inventory, the DPRK would see UAS as a more economical and less sophisticated alternative to qualitatively and quantitatively conduct aerial ISTAR, as well as some attack operations including the delivery of chemical and biological agents. Hence, although the DPRK's UAS are still nascent, further modernization and production is likely given their effectiveness for asymmetric attacks and reconnaissance.

Given the cost-intensive nature of aerial platforms, the DPRK has struggled to update the KPAAF inventory. The issue is not simply about the DPRK's inability to acquire next-generation platforms but also the shortage of platforms that would be essential for the KPAAF's missions. For instance, even during the Cold War, the KPAAF did not acquire the MiG-25 that boasts high climb rates and capability to operate in significantly high altitudes, making them suitable for interception and reconnaissance missions. The DPRK reportedly made attempts to acquire modern aircraft such as the Su-30 and Su-35 from Russia as well as the J-10 and J-11 from China but were turned down.[68] As long as the DPRK struggles to acquire more advanced tactical aircraft, the KPAAF is constrained to focus on sustaining the airworthiness of its limited aircraft and making upgrades to its anti-air defense systems.

Weapons of Mass Destruction

The most prominent development in the KPA is the production of WMD including ballistic missiles and chemical, biological, radiological, nuclear, and explosives (CBRNE) arsenals. For the DPRK, WMD are useful not only for their strategic effect but also in their tactical applications to inflict destruction or disruption on the battlefield. As explained in Chapter 5, the DPRK began pursuing the development of strategic weapons in the 1960s, and their efforts came to fruition in the 1990s and 2000s when Pyongyang began conducting tests of various ballistic missiles and nuclear devices. More significant developments unfolded in the Kim Jong-un era, demonstrating significant technological advances as well as the exponential increase in the frequency of ballistic missile and nuclear tests. Although Kim Jong-un declared a moratorium on the testing of ICBM and underground nuclear tests in April 2018, the DPRK has continued to develop its ballistic missile inventory as demonstrated at the October 2020 military parade.

Nuclear weapons

Although the DPRK started their nuclear weapons program in the 1960s, little was known about the actual progress until the 2000s. To date, the DPRK has conducted six underground nuclear tests with the first taking place on 9 October 2006. Tests to date have revealed the growing lethality of the DPRK's nuclear arsenals with the steady growth in the explosive yield. In particular, the test on 3 September 2017 was claimed by some of having a yield of 250kt, suggesting that the DPRK may have produced a hydrogen bomb.[69] While the explosive yield from the DPRK's tests may not compare to the USSR's testing of the RDS-220 hydrogen bomb in 1961 that recorded 50,000kt, or even the tests conducted by China, France, the United Kingdom, or the US, it is still approximately seven to ten times larger than the arsenals dropped on Hiroshima and Nagasaki in August 1945.

There are significant challenges in measuring the quantity and quality of the DPRK's nuclear arsenals, as well as their capacity to further build their stockpiles. Estimates have varied greatly, ranging from 20 up to 50 (or more) nuclear warheads. The key question is whether the DPRK's nuclear arsenals are composed of plutonium or uranium. Following his visit to the Nyongbyon facility in November 2010, Siegfried Hecker (former director of the Los Alamos National Laboratory) stated that the DPRK's nuclear technology has advanced to the stage where it can enrich both plutonium and uranium.[70] If Pyongyang is capable of enriching uranium, then that will allow them to build their nuclear weapons stockpile at a much faster rate. Hence although the DPRK's nuclear weapons stockpiles are still small compared to most other nuclear weapons states, much attention is required on how Pyongyang will continue its efforts for production, as well as operationalizing the capabilities.

One must keep in mind that the nuclear threat posed by the DPRK is not only based on the quantity or quality of the arsenals they possess but the production capacity. Pyongyang's moratorium on underground nuclear tests and the closure of the Punggye-ri Nuclear Test Site suggests in no way, the reduction of its nuclear weapons capabilities. Rather, reports have claimed that the DPRK is in fact upgrading the facilities at the Nyongbyon Nuclear Scientific Research Center as well as at least two clandestine enrichment facilities.[71] While the existence of these facilities contradicts the mood created from the diplomatic interactions in 2018, they are nonetheless consistent with Kim Jong-un's New Year declaration made in January 2018 for the mass production of nuclear arsenals. Thus, as long as the DPRK has operational nuclear facilities at hand, the production of nuclear weapons would nonetheless continue.

The other major concern is electromagnetic pulse (EMP). Even if the DPRK does not (or cannot) deliver a nuclear warhead to its intended target, it could set off an EMP blast that causes indirect lethal effects. An EMP strike on a city would wreak havoc by incapacitating and disrupting critical infrastructures including energy, finances, information, navigation, communications, health and medical services, and transport and logistics. Militarily, the dependence of the US, ROK, and Japanese forces on C4ISTAR and other digitalized systems make them vulnerable to EMP strikes, serving as opportunities for the DPRK to narrow the technological gap and possibly even turning the tables to make tactical gains.

The DPRK could also stage nuclear attacks without ballistic missiles – particularly against closer targets in South Korea. The KPA is capable of tactical strikes through nuclear-tipped munitions using artillery and MLRS, aerial strikes using swarm tactics as well as deploying special operations units. For instance, reports claim that the selected number of elite soldiers from reconnaissance platoons and light infantry brigades in the 7th and 9th Corps is being armed with nuclear backpacks to spray radioactive material.[72] In many regards, nuclear attacks using the earlier methods are easier and in some cases more reliable for attacks against the military and civilian assets in South Korea.

Chemical and biological weapons

For the DPRK, chemical and biological weapons serve as important assets to gain an edge in several ways. First, chemical and biological weapons allow the DPRK to incapacitate opposing forces and populace without physically destructing infrastructures, thereby allowing the KPA to capture and exploit those assets. Second, preventative and countermeasures against chemical and biological weapons seldom keep pace with the actual threats, allowing the KPA to undermine the opponents' defense and level the playing field. Third, chemical and biological weapons are psychologically effective, triggering fear and panic among the government and citizens that are vital conditions for the DPRK's hybrid warfare.

Although the DPRK is a signatory of the Geneva Convention that bans the use of chemical weapons in war, they have not yet acceded to the Chemical Weapons Convention. According to the ROK Ministry of National Defense, the DPRK is suspected to possess approximately 2,500 to 5,000 tons of chemical agents such as mustard blister agents and nerve gas (e.g., Sarin and VX).[73] Reports suggest that the DPRK attained its indigenous capacity to produce a variety of chemicals while also importing various dual-use precursors.[74]

Regarding biological weapons, despite being a signatory to the 1987 Biological Weapons Convention and the Geneva Protocol, the DPRK is believed to have a significant stockpile of biological weapons that have been developed since the 1960s.[75] Studies claim that the DPRK has at least 13 types of biological agents including Alimentary Toxic Aleukia, Anthrax, Botulism, Brucellosis, Cholera, Dysentery, Korean Hemorrhagic Fever, Plague, Smallpox, Staph, Typhoid Fever, Typhus Fever, and Yellow Fever.[76]

There are difficulties in quantifying the DPRK's chemical and biological weapons stockpiles, particularly with the notable differences in estimates by government and non-government institutions or testimonials by officials and defectors. A number of reports claim that the DPRK tests its chemical and biological agents on political prisoners. Yet such actions only reflect the cruelty of the regime rather than the actual effectiveness of the weapons themselves. Rather, the question is how the DPRK can effectively deliver the chemical and biological weapons.

Although chemical and biological agents are easy to miniaturize and weaponize, their delicate nature limit the means of delivery. Chemical weapons could be delivered via ordnances including missiles as well as personnel but with specific means to prevent the agents from unintentional dispersal, evaporation, and incineration. As for biological weapons, given that they are mostly living pathogens make them implausible to deliver via missiles or other ordnances as the agents could simply burn up. Moreover, both chemical and biological attacks are only effective in an environment where there can be direct contact (for chemical weapons) or a contagious effect (for biological weapons). Thus, essentially, the most effective method of delivering chemical and biological weapons would be either through direct contact and dispersal by personnel, animals, low-flying aircraft or transmission through infected objects and surfaces – although this also requires entering enemy territory.

The aforementioned problems do not mean that the DPRK's chemical and biological weapons are harmless. Above all, civilians are less protected than military, civil defense, and law enforcement personnel, making them highly vulnerable to chemical and biological arsenals. Chemical weapons could be delivered covertly for both large-scale indiscriminate attacks and assassinations such as the case with the Tokyo subway Sarin attack by Aum Shinrikyo in March 1995 and the assassination of Kim Jong-nam with VX at Kuala Lumpur International Airport in February 2017. As for biological weapons, the effects of diseases have been proven by various epidemics and pandemics, most notably with COVID-19. Thus, the highly lethal and wide-spread effects of chemical and biological weapons still make them one of the most threatening weapons of the KPA.

Ballistic missile program

Ballistic missiles are arguably the most modern weapons of the KPA and also the only means of power projection. For over five decades, the DPRK devoted much effort into building and diversifying its ballistic missile inventory that are currently commanded by the KPASRF. Although the DPRK conducted a number of test launches from the 1980s until the 2000s, much of them failed. Yet entering the 2010s, the DPRK has made significant advancements evidenced by not only the greater number of successful launch and engine tests but also major developments in mobile launch vehicles including SSB that allows the KPA to conduct surprise attacks and second-strikes.

The most powerful is the liquid-fueled ICBM, most notably with the Hwasong-13, Hwasong-14, Hwasong-15, and the KN-14 that all have the potential to hit much of the continental US.[77] More recently at the 75th anniversary of the WPK parade, the DPRK unveiled a monstrous new ICBM larger than the Hwasong-15 with presumably longer range and larger payload. As for IRBM that is capable of striking Japan and Guam, the DPRK has largely learned from the Soviet R-27 SLBM. The Musudan BM-25 was first unveiled at the 75th anniversary of the KPA in April 2007 but suffered a series of failures in the test launches in 2016. The successor to the Musudan BM-25 is the Hwasong-12 that has been much more successful in its test launches and with much more powerful specifications.[78]

The DPRK's MRBM came into service in the mid-1990s with the Nodong-1 (also claimed to be an upgraded version of the Hwasong-6 SCUD-C SRBM) and the Hwasong-9 – both of which are single-stage liquid propellant.[79] In recent years, the DPRK has worked on the Pukguksong-2, a two-stage, solid propellant MRBM with much greater range than previous models.[80]

The DPRK also continues to build its SRBM inventory that are capable of striking high-value targets in South Korea. From the mid-1980s to the early 1990s, the DPRK introduced two SCUD-based missiles known as the Hwasong-5 and Hwasong-6.[81] The more contemporary SRBM have been more diverse. In 2019, the DPRK has tested three types of single-stage solid fuel SRBM including: the KN-23 that closely resembles Russia's solid-propellant Iskander-M (9K720) and the ROK's Hyunmoo-2B; the KN-24 with lower trajectory to dodge missile defense systems; and the KN-25 designed to be launched from a multi-tube TEL.[82] Moreover, the KN-18 first tested in May 2017 seems to be a maneuverable reentry vehicle (MaRV) variant of the SCUD that could be used against carrier strike groups.[83]

In the 2010s, the DPRK ventured into SLBM capabilities to gain capabilities for surprise attacks and second-strike. In 2014, the DPRK was reported to have conducted a number of static ejection tests. Then in May 2015, the DPRK conducted the first publicized launch of the Pukguksong-1 followed by five more over the next 13 months that were mostly unsuccessful. It is widely believed, however, that the Pukguksong-1 was not launched from a submarine but a submergible barge in Sinpho.[84] The nature of the missile and the tests suggested that the Pukguksong-1 was a technological demonstrator. After some time of silence, the DPRK conducted the first cold-launch test of the Pukguksong-3 in October 2019 that is estimated to have much longer range.[85] The DPRK showcased the Pukguksong-4 in October 2020 and the Pukguksong-5 in January 2021 with unknown, but presumably more advanced specifications. Still, there are nevertheless numerous hurdles going forward, including the operationalization of a SSB/SSBN and the ability to reliably launch SLBMs from the vessels.[86]

One notable development in recent years is the shift from fixed launch facilities to more mobile methods using MEL, TE, and TEL as well as SSB/SSBN. Mobile platforms not only allow the DPRK to disperse, conceal, and bunker (or submerge for SSB/SSBN) the missiles for surprise attacks and second-strike but also hide and protect them from preventative, preemptive, and counter-strikes. The DPRK is known to have acquired and developed a variety of MEL, TE, and TEL since the 1980s, with much of them based on vehicles acquired from China and the USSR/Russia.[87] In recent years, however, the DPRK seems to have become capable of producing its own TEL, with the unveiling of a 9-axel vehicle at the test launch of the Hwasong-15 in November 2017 and the enormous 11-axel vehicle paraded at the 75th anniversary of the WPK in October 2020.

Despite the progress in the DPRK's ballistic missile program, several questions remain on whether they can reliably deliver warheads to its intended target. The first concerns accuracy (or the circular error probability), with questions over whether the DPRK has shifted from early guidance to satellite guidance systems. In the early years of its ballistic missile development, the DPRK focused more on range than accuracy. For instance, Choi Ju-hwal, a former Colonel of the KPA who defected to the ROK, stated that DPRK missiles are designed to make an impact in a "general region" as opposed to a pinpoint surgical strike.[88] Still, precision will be critical for missiles targeting specific high-value targets and moving targets such as aircraft carriers.

Second, there are continued debates over whether the DPRK's nuclear warheads are miniaturized enough to be mounted onto ballistic missiles. From the early 2010s, there were growing speculations that the DPRK has cleared the technological bar. The North Korean state media also claimed that the underground nuclear test on 12 February 2013 was a detonation of a miniaturized device. On a number of occasions, the DPRK has also publicized photographs of Kim Jong-un posing with nuclear warheads that seem (at least by appearance) compact enough to be fitted onto ballistic missiles. Still, even if the DPRK can produce miniaturized warheads, there are still questions over whether the device is operational and functionally effective.[89]

Third, there are questions over the survivability of the reentry vehicle, terminal-stage guidance, and warhead activation. The North Korean state media has claimed its technological milestones on numerous occasions. On 9 March 2016, the North Korean state media revealed photographs of a circular nuclear warhead to be fitted onto a reentry vehicle. Then on 15 March 2016, the DPRK conducted a test of a reentry vehicle, exposing it to the exhaust of a missile engine to simulate the heat and pressure similar to reentering the earth's atmosphere. The tests, however, are only partially indicative of the actual capabilities. The flight test of the Hwasong-14 on 28 July 2017 proved otherwise, with the reentry vehicle disintegrating roughly at 3,000 to 4,000 m in altitude 200 km off the western coast of Hokkaido in Japan.[90] Although the DPRK could have indeed expected the reentry vehicle to burn up and use the data for further studies, the test gave credence to the conclusion that the DPRK's ballistic missiles technologies are still incomplete.

Going forward, the DPRK seeks to develop more solid fuel long-range ballistic missiles including SLBM, multiple independently targetable reentry vehicles (MIRV) warhead, as well as hypersonic weapons to enhance their strategic and tactical strike capabilities.[91] Yet whether the technological progresses can be collated into credible levels of readiness remains unclear unless the DPRK actually conducts an armed launch such as an atmospheric test or an actual strike. Still, the technological hurdles are unlikely to deter the DPRK, and the military parade in October 2020 evidenced the DPRK's unwavering will to advance its ballistic missile capabilities.

Cyber warfare

In recent years, the DPRK has also put much effort into enhancing its cyber warfare capabilities for both offensive and defensive operations, including psychological warfare, denial of service (DoS), distributed denial of service (DDoS), hacking, espionage, and other operations.[92] The advancements in the DPRK's cyber warfare capabilities largely owe to the developments in ICT since the 1990s. By 1998, the DPRK established Unit 121 (now within the RGB) specialized in cyber warfare. Moreover, reports claim that some of the cyber warfare units are deployed to locations outside of the DPRK, with activities conducted in areas such as Shenyang in China.[93] Cyber-attacks are of grave concern in the fourth-industrial revolution era given the integration and dependence of critical infrastructures and services in cyberspace. Furthermore, cyber-attacks can also be kinetic. For instance, cyber-attacks that cause an abrupt shutdown of a nuclear reactor would lead to a meltdown that would result in major damages and casualties.

Assessing the DPRK's actual readiness for cyber warfare is difficult given their secretive nature. At this stage, much of the DPRK's cyber-attacks have focused on crippling and disrupting government and corporate systems as well as exploiting social media and spreading fake news for psychological warfare. Since 2009, the DPRK is alleged to have conducted several major DDoS attacks on the US and ROK government and civilian critical infrastructures as well as the media. In this regard, the operations were more of an irritant, as opposed to inflicting detrimental effects such as critically shutting down critical infrastructures or causing kinetic incidents. Still, there are credible concerns about the DPRK's ability to conduct more lethal operations that target critical infrastructures and information networks that are vital for modern militaries. Some such as Lewis has suggested that the DPRK's cyber capabilities have only proved to be successful against poorly guarded sites and do not "duplicate the effect produced by Stuxnet or the Russian attack on a Ukrainian power facility."[94] While that may be the case for now, the hurdles in developing more sophisticated and lethal means of cyber-attacks are likely to be far less expensive and sophisticated.

Although the DPRK's cyber warfare capabilities are threatening, they seem to have critical weaknesses in defensive cyberspace operations. In December 2014, the information networks in the DPRK were shutdown, prompting Pyongyang to blame the US. Whether it was the US or hacktivists that was responsible for paralyzing the DPRK's information network is beside the point. Rather, the significance lies in the fact that the limited and isolated nature of the DPRK's cyber systems make them extremely vulnerable.

Electronic warfare

Electronic warfare capabilities have been a vital means for the KPA in tilting the force-on-force balance by disrupting and degrading the C4ISTAR capabilities of the US, ROK, and Japanese forces. The DPRK's ambitions to develop electronic warfare capabilities came about in the 1960s but accelerated in the 1990s after witnessing the advanced network-centric systems of the US forces in the Gulf War, leading to the establishment of various academies, research programs, and technological acquisitions.[95] Electronic warfare is well systemized in the KPA, with the Classified Information Bureau, Communications Bureau, Electronic

Warfare Bureau, and RGB of the MND, and in almost every command or corps of the KPA that consist electronic warfare units.[96]

In hardware, the KPA has GPS and satellite jammers below the Pyongyang–Wonsan line, and some platforms particularly in the KPAAF have been retrofitted with electronic warfare and/or electronic warfare support capabilities.[97] While the exact specifications of the DPRK's electronic warfare assets are unknown, their capabilities have proved effective to a certain extent. According to reports, the series of GPS jamming by the DPRK affected approximately 1,137 aircraft and 265 ships since 2010.[98] Like the cyber warfare operations, much of the activities by the DPRK were either to harass the ROK and US or responses against their military exercises.

Questions remain over the KPA's abilities in the context of electronic warfare protection and support. Indeed, the KPA assets may not be as vulnerable as the US, ROK, and Japanese counterparts who are much more dependent on network-centric systems. Nevertheless, the KPA has become increasingly vulnerable to disruptions and interceptions over the years as they advanced in their C4ISTAR assets and procedures. The DPRK is clearly concerned, with field manuals emphasizing the importance of using radar-absorbing paint and other concealing techniques.[99] In hardware, the KPA has made some electronic countermeasure (ECM) resistant modifications including the employment of chaffs and flares.[100] Considerable efforts are also seen in communications security, including the relocation or fortification of assets, frequent changing of codes, and minimizing the use of communication systems. Yet beyond the rudimentary techniques, how the DPRK can acquire and operate its electronic warfare protection systems are questionable given the lack of access to technology.

C4ISTAR

One critical area that needs assessment is C4ISTAR systems that are pivotal components of modern militaries. Advancements in ICT have not only led to significant improvements in the ability to detect, locate, track, target, and strike opponents but also to work with a common operational picture to effectively and efficiently execute operations. In the case of the DPRK, despite some notable developments in the increased levels of computerization and development of network systems, they still suffer from shortages in C4ISTAR assets for better forms of real-time operational command and control.

In modern times, technological advancements have significantly enhanced all forms of intelligence gathering, including human intelligence (HUMINT), open source intelligence (OSINT), signals intelligence (SIGINT), imagery intelligence (IMINT), and measurement and signatures intelligence (MASINT). Combined, the various forms of intelligence have allowed states with both timely strategic and tactical intelligence to effectively and efficiently operate its military forces. Indeed, one of the KPA's strengths stems from the long-lasting deadlock on the Korean peninsula, where they have gained familiarity with the ROK and US assets, as well as the geographic characteristics of the Korean peninsula and surrounding areas. Referring to the DPRK's coordinated, timely interception of the USAF RC-135S in 2003, Scobell and Sanford note Pyongyang's "considerable degree of pre-mission intelligence collection and planning."[101] Yet despite the advantages, there are severe deficiencies in the KPA's C4ISTAR capabilities – particularly with those that involve sophisticated technologies.

For the DPRK, HUMINT has been one of the key strengths with the high numbers of agents and pro-DPRK sympathizers conducting intelligence collection, counterintelligence, and psychological operations. In the past, a number of agents involved in espionage and conspiracy have been nabbed in South Korea and Japan. In the 1970s and 1980s, North Korean operatives had abducted foreign citizens for intelligence and to provide cultural and language training for spies. In another example, the ROK was rocked by the "Korean mata hari" scandal in 2008 that involved a DPRK operative known as Won Jeong-hwa who had sexual relations with ROK Army officers to extract confidential military information.

The DPRK also has credible capabilities in OSINT, obtaining information either through officials and secret operatives abroad or simply via online to study the white papers, budgets, guidelines, and other unclassified and declassified documents as well as news, and social media to decipher as much information as possible about the developments and policies of the states of concern.

While the DPRK's strengths in HUMINT and OSINT are useful for strategic decisions and psychological operations, there are limitations in the context of real-time tactical applications. Real-time ISTAR are essentially gained through IMINT, MASINT, and SIGINT. On the ground, the DPRK can gain images captured by personnel deployed near the MDL, as well as operatives based in South Korea and abroad. The KPAN and the RGB also operates various vessels for ISTAR operations in littoral areas albeit with limitations in the quality and quantity of data gathered. As for the KPAAF, reports have suggested that the selected number of the An-24, Il-28, Il-76 as well as older tactical combat aircraft have been converted AEW/AWACS applications.[102] The use of civilian aircraft for ISTAR is also possible, although Air Koryo barely ever flies into South Korean, Japanese, or US territory, constraining them from having any real effect. In recent years, the DPRK has also developed drones fitted with digital cameras presumably to gather photo imagery of key critical infrastructures in South Korea, although the level of technological sophistication seems to be rudimentary and questionable in their effectiveness.

Obviously, the effectiveness of rudimentary and makeshift platforms is limited. Yet the greatest problems lie not in the aircraft, vehicles, and vessels but the deficiencies in the radars and sensors. Like the aforementioned sonar systems, the vast majority of the radar and sensor systems in the KPA are those acquired from China and the USSR during the Cold War that are technologically marginal by today's standards. Most of the early warning and ground control/intercept radars are those from the 1950s, including Back Net, Backtrap, Bar Lock, Big Back, Flat Face, Knife Rest, Moon Face, Spoon Rest, Squat Eye, Tall King, and Tin Shield.[103] Even for fire control and guidance/tracking radars, many are those of the 1960s and 1970s, including Drum Tilt, Fan Song, Fire Can, Gin Sling, Low Blow, and Square Pair.[104]

In addition, the DPRK is known to have two types of height finding radars being Odd Pair and Side Net.[105]

Despite the problems in radar and sonar systems, developments in ICT have led to some developments in the KPA's C4ISTAR assets. Until the 2000s, the developments in communication systems in the DPRK focused on the systemization of landlines, radio, and microwave transmission systems as well as pagers.[106] The implementation of a C4ISTAR system in the KPA during this time was incremental or partial at best. For example, Bermudez noted in 2005 that the DPRK operates its own air and coastal defense networks, although the level of computerization was claimed to be low.[107] Notable levels of computerization have taken place since, as seen in the equipment at the National Aerospace Development Administration Satellite Control Centre and various strategic and tactical command and control facilities, as well as in Kim Jong-un's field guidance and inspection tours of military exercises. Pyongyang also seems to have built their own tactical C4ISTAR equipment including mobile gadgets. Moreover, Glocom, a company operated by the RGB, was reported to have marketed some tactical C4ISTAR systems, suggesting that the KPA operates the same or more advanced equipment.[108]

Regarding satellites, the DPRK has pursued the development of its own satellite network but with limited success. On 31 August 1998, the DPRK launched the Kwangmyongsong-1 followed by the Kwangmyongsong-2 on 5 April 2009, and Kwangmyongsong-3 on 13 April 2012 – all of which failed. It was not until 12 December 2012 when the DPRK was able to put a satellite into orbit with the launch of Kwangmyongsong-3 followed by the Kwangmyongsong-4 on 7 February 2016. The DPRK has claimed to be working on developing its own reconnaissance satellites, although the actual progress remains unknown, and it is assumed that Pyongyang would remain dependent on foreign systems in the meantime. Today, there are four satellite navigation systems, including the US-led GPS, the European Union's Galileo, Russia's GLONASS, and China's Beidou, with the latter two being logical choices for the DPRK. Although the DPRK may well have utilized GLONASS in the past, recent developments indicate that Pyongyang is utilizing Beidou.[109] In July 2014, a delegation of North Korean engineers received training at the National Remote Sensing Center run by the Chinese Ministry of Science and Technology. Given the fast advancing nature of Beidou, there are genuine concerns over how this could enhance the KPA's C4ISTAR and ballistic missile capabilities.

Yet despite the developments, fully installing and operating modern C4ISTAR systems are far from easy. Above all, there is the cost factor, not only with the C4ISTAR technologies themselves but also the need to configure and upgrade the KPA's existing installations and platforms so that those systems can be properly operated and networked. Moreover, the full implementation of C4ISTAR systems would involve organizational and procedural changes including decentralized control and horizontal communication, raising issues for the regime that has been adamant about maintaining absolute command and control. Even if the aforementioned hurdles are overcome, the technological gaps between the KPA and the US, ROK, and Japanese counterparts are apt to remain large.

Operational readiness

Operational readiness makes or breaks a state's overall military readiness. Even if a military possesses a robust force structure, deficiencies in organizational management, logistics, maintenance, supplies, personnel, as well as education and training would significantly undermine its overall readiness. Hence, the critical question is whether the KPA is adequately prepared for military operations especially under war conditions.

Supplies and logistics

Arguably, the biggest problem in the KPA's operational readiness is the deficiencies in energy, food, munitions, and also the overall logistical supply chain. Despite the high allocation of resources to the military sector, they have proved to be insufficient in feeding and fueling the KPA's massive force structure, particularly when one looks at how many of the KPA units are forced to undertake labor to self-sufficiently secure their own food and finances.

Greater problems are seen in wartime supplies, where the DPRK is said to have only two to three months of ammunition, food, petroleum, oil, and lubricants.[110] Yet the critical point is not simply the quantity of the supplies but also poor management. Although Pyongyang has established various depots within its territory, the quality and quantity of reserve stockpiles have declined due to decay. Moreover, despite the supposed strict oversight of supplies, various reports claim that food rations in many installations have been subject to theft by desperate personnel.

There are also questions regarding the quantity of munitions that would be one of the key determinants to the sustainability of the KPA's firepower. While it is widely believed that the DPRK has a large stockpile of munitions for war, it is questionable whether the stockpiles are sufficient. For instance, if the DPRK does attempt to turn Seoul into the "sea of fire," this will force the KPA to consume overwhelming amounts of shells and other munitions to neutralize the ROK and US forces near the MDL as well as striking Seoul and nearby cities. Yet this is merely the first challenge. Even if the DPRK manages to achieve the above, the KPA will then need even more munitions to make further advances and engage the counterattacking ROK and US forces.

Arguably the most critical logistical problem for the KPA is the chronic energy shortage that limits the KPA's operational range. One of the key constraints to the DPRK's air and naval power projection comes from not only the shortage of fuel but also the lack of refueling platforms that are vital in improving the efficiency of operations by extending the tactical units' range and reducing the number of deployments. Currently in the KPAAF, the only possibility is the conversion of Il-62 or Il-76 transport aircraft, although retrofitting such platforms would nevertheless be costly and questionable in their actual effectiveness. Of course, one could argue that refueling capabilities are of low priority for the DPRK given that they do not envision fighting in areas beyond their immediate periphery. Nevertheless, the KPA air and naval units' limited operational radius inevitably impair their anti-access and area-denial capabilities, consequently making them more vulnerable.

Certainly, the DPRK leadership's avoidance of actively investing and relying on fuel-intensive tactical aircraft and heavy-duty naval platforms eased the O&M costs to some extent. Nevertheless, the DPRK is not completely free from O&M burdens due to the obsolescent nature of the KPA installations and platforms that are less fuel-efficient causing consumption rates to increase. Given the size of the KPA, the amount of fuel required to make the whole force operational is astronomically high. Although Pyongyang has historically depended much on Beijing and Moscow for gas, oil, and petroleum, the provisions have been far from sufficient. Moreover, the series of economic sanctions vis-à-vis the DPRK seems to have had an impact on the KPA. For instance, Asia Press reported that economic sanctions have caused gas prices to skyrocket, forcing some units in the KPA to use bullock carts and wood-combusted vehicles to transport goods.[111] Consequently, the little available fuel is saved for the most vital operations or units, undermining the overall functionality of the KPA. Hence, even if the DPRK manages to secure sufficient amounts of energy, considerable amount of time and configurations are needed before these platforms can be revived to full operational status.

Although management is to blame for the fuel shortages, the problems have not necessarily brought the KPA to its knees. Despite remaining to be dependent on its benefactors, the DPRK has tried to be as self-reliant as they can in operating the capabilities they possess. A prime example is the DPRK's efforts to produce its own rocket fuel. In 2017, a number of observers claimed that the DPRK has the capability to produce unsymmetrical dimethylhydrazine – a fuel used in ballistic missiles. The DPRK's effort to indigenously produce the fuel is hardly a surprise given that they have pursued advancements in heavy chemical industry and natural sciences for military applications over many decades.[112] Rather, the question is whether the DPRK's indigenously produced rocket fuel is both qualitatively and quantitatively sufficient to operate all of the KPA's ballistic missiles.

Critical problems are also seen in transportation.[113] Although the KPA possesses and has access to the best vehicles and transport infrastructures including civilian assets, severe problems are caused by not only energy shortages but also the state of infrastructures. On the ground, much of the DPRK's roads are unpaved and the extensive tunnel networks have disadvantages in terms of traffic flow. In rail, approximately 98% of the railroads in North Korea are claimed to be single-track and poorly maintained.[114] Even in maritime transportation, the DPRK did attempt to construct an east–west canal in the 1970s to connect the Taedong and Kumya rivers, yet the project fell far short of completion.[115] Severe problems are also seen in aviation sector, where only a portion of the airbases are fully operable or have paved runways. Making matters more challenging, the DPRK's transport systems lack intermodality that severely undermines the KPA's logistical efficiency.

Issues in transportation are also created by the KPA's operational art. The DPRK's military logistical supply chain has been much dependent on secrecy to conceal the location of facilities and also the operations of the KPA. KPA transport units often operate at night or during times of low visibility, and the entry and exit points to the facilities are often hidden or connected in deceptive fashion. While effective to some extent, the major downside is that such practices and structures undermine smooth transportation of goods and personnel, consequently affecting the KPA's readiness.

The problems in transportation raise issues for wartime operations. Indeed, the DPRK has experience in overcoming transportation issues, such as the use of K-6 S-type pontoon sets and underwater bridges during the Korean War.[116] Nevertheless, there are limits to the applicability of past techniques to present and future operations, given that the KPA is much larger and mechanized and operate under different conditions. Training would be the most viable prescription to work effectively under new conditions, and the KPA could theoretically adapt to operating on rough terrains. Still, there would be outstanding challenges, particularly for operators of heavy-duty platforms who are already handicapped by the compromised training aids and lack of fuel.

Maintenance

The outdated nature of the KPA inventory reflects not only the DPRK's problems in acquiring platforms but also their troubles in maintaining and repairing them. While small-scale repairs to equipment and platforms can be handled in bases, major repairs to heavy-duty platforms take place in larger bases or factories. Without doubt, one major factor in why the DPRK was so adamant about establishing its own military–industrial complex was to produce parts and conduct repairs. However, the ability to conduct large-scale maintenance and repairs hinges on capacity. Even if the DPRK has managed to keep old platforms in workable order, the combination of economic problems and lacking access to technology suggest that the KPA faces grave problems when the demands for maintenance and repairs exceed its already limited capacity.

The KPA's problems in maintenance and repairs are reflective of the state's engineering capacity and access to technology. Given that much of the DPRK's indigenous weapons systems are based on various Chinese and Soviet/Russian imports, the DPRK presumably would have worked to secure stockpiles of spare parts and technological knowledge. However, the sufficiency of the stockpiles is questionable. While it is true that some parts could be produced domestically, some critical parts are only available in the platform's country of origin. Problems in acquiring components are not only due to cost and reluctance of overseas suppliers but also because they are simply no longer available.[117]

Consequently, the problems in engineering capacity and access to technology have forced the DPRK to conduct rudimentary maintenance and repairs. Cannibalism has become common practice in the KPA, with decommissioned platforms becoming sources of spare parts for operational units.[118] While reusing parts from obsolete platforms could be a cost-saving alternative, the dilemmas in ensuring optimal readiness of the KPA platforms are clear. On the one hand, continuous acquisition of new platforms will require either the mass import or production of spare parts. On the other hand, the continued use of dated platforms means that the frequency of maintenance and repairs would increase in congruence with the platforms' age due to wear and tear. Both options are pricey, and the reality suggests that there is no easy way for the DPRK to keep the KPA platforms in optimal shape.

Personnel

Currently, the KPA has approximately 1.28 million active personnel while there are approximately 189,000 active personnel in the KPISF.[119] As for reserves, the RMTU has approximately 600,000 while the WPRG has approximately 5.7 million, and the RYG has roughly 1 million personnel.[120] Given that the North Korean population is estimated at 25 million, almost a third of the population serves in the KPA and reserve/paramilitary units. Yet despite the large pool of personnel, the level of readiness has been undermined by the state's socio-economic problems particularly with the myriad economic, health, and demographic issues since the "arduous march" of the 1990s.

The socio-economic circumstances have severely impacted the health and morale of personnel. The health problems in the KPA have been serious, where even the elite or more privileged personnel are also suffering from myriad health problems. In a recent case, a KPA soldier who fled to South Korea over the MDL on 13 November 2017 was found with hepatitis B and parasitic worms. Given the situation even among the more elite and privileged personnel, one can presume that the situation would be worse in the rear units where the logistical shortfalls are direst. Even for morale, despite the intense ideological indoctrination, personnel complaining about the leader are not uncommon, and defections by KPA personnel are no longer rare. Moreover, there is considerable level of favoritism and nepotism in the KPA, with draftees from elite families given greater flexibility in their appointed positions.[121] Recruits are known to prefer easier jobs, or those that may bring benefits from bribes (e.g., border guards or internal security units). Deteriorating morale in the KPA has also led to misconduct including assault, bribery, embezzlement, extortion, murder, racketeering, rape, robbery, and theft. In recent years, the private markets have also lured KPA personnel into becoming not only consumers but also suppliers, selling goods stolen from military reserves.[122]

Major compliance problems are also seen in information security – one of the areas the DPRK claims as one of the most vital to their military power.[123] For many years, the secretive nature of the KPA was one of the DPRK's advantages. Yet in recent years, intelligence leaks from the DPRK has surged, with an increasing number of KPA personnel becoming involved in the selling of military information to foreign media and even governments. Furthermore, the defection of

KPA personnel has also increased in numbers, with many having either worked in or possessing contacts in the government or the military. Of course, leaked information is not guaranteed in their authenticity and could be outdated, fabricated, or simply inaccurate. Nevertheless, corrosions in the protection of information would expose features of the KPA that undermine their strengths and expose their vulnerabilities.

There are also questions regarding the impact of generational changes on the KPA. Until the 1990s, the KPA was led by battle-hardened veterans of anti-Japanese partisan campaigns or the Korean War such as Choe Hyon, Choe Kwang, Jo Myong-rok, Kim Chol-man, O Jin-u, Ri Ul-sol, and Ri Yong-mu. Since the 2000s, the top echelons of the KPA are those who played key roles during the latter decades of the Cold War, such as Kim Il-chol, Kim Jong-gak, Kim Yong-chol, Kim Yong-chun, O Kuk-ryol, Ri Myong-su, and Ri Yong-ho. While many of the officers who built their careers during the Cold War are hardliners, their battlefield experience is based on the small-scale attacks and provocations rather than war. In the future, there will be mixed results as the younger generations join the senior ranks of the KPA. Unlike the current generation of senior cadres in the KPA, the younger generations have spent many more years under Kim Jong-il and Kim Jong-un than Kim Il-sung, and grew up and lived through the socio-economic crises of the 1990s and 2000s. Moreover, the younger generations in the KPA also grew up and worked through the years when the military was not only acquiring a variety of new weaponry but also witnessed the deterioration of discipline and morale among the personnel. It is hard to forecast how the generational changes impact the KPA. On the one hand, the effect may be limited given the centralized and politicized nature of the command and control system, and the fact that senior military cadres come from elite backgrounds that make them loyal to the leader by default. On the other hand, however, the younger generations could also advise the leader with views that are different from the older generations of cadres.

For the DPRK, the options to solve the human resource issues in the KPA are few and far between. One obvious prescription would be to increase the pay, provide more food, and also ensure better distribution of equipment and weaponry, although such measures require significant increases, or at least reconfiguration of the already over-stretched military budget. Alternatively, Pyongyang could refine the military's personnel numbers, but doing so would undermine the quantitative strengths, and cause imbalances in the composition of the KPA. Both measures, however, will still fall short of remedying the socio-economic issues that gravely impact the KPA for the long-term and also the military's overall readiness.

Education and training

Like many militaries, education and training are pivotal to the KPA's readiness and integrity. To date, the content of education and training in the KPA has been consistent in the way that they have taught the long-standing warfighting concepts and doctrines as well as, of course, political loyalty to the leadership. At the same time, however, there has been notable changes in the KPA's education and training systems and processes due to not only institutional and policy developments but also circumstances.

The establishment of the KPA in 1948 and the developments under the Line of Self-Reliant Defence involved the large-scale institutionalization of education and training. Officer candidates study in academies according to their services, such as the Kang Kon Military Academy, Combined Artillery Officer School, and Armor Officer School for the army; Kim Jong-suk Naval Academy, and Naval Officers School for the KPAN; and Kim Chaek University of Air Force, and Kyongsong Flight Officers School for the KPAAF. There are also a variety of specialized academies, such as the Pyongyang Computer Technology University, Kim Il Military College (formerly Mirim College), O Jin-u Artillery Academy, as well as those specialized in training operatives and political officers. Senior officers receive training at Kim Il-sung Military University, Kim Il-sung University of Politics, Kim Jong-il Military Postgraduate Institute, and the Kang Kon Military Academy. In addition to the elite schools, the various commands and corps of the KPA have tactical training institutions for officers and non-commissioned officers such as the Naval Technical Training Center.[124] The array of educational and training institutions allows not only the nurturing of expertise but also the formulation of tactical concepts and doctrines that is vital to enhance the KPA's readiness.

Assessing the KPA's training regimes is hard given the nature of the North Korean state media. Occasionally, the North Korean state media gives coverage of KPA drills that supposedly simulate amphibious landings, anti-access operations, bombardment of targets, or raids on ROK facilities such as the mock Blue House constructed southeast of Pyongyang. Yet there is much less than meets the eye in such orchestrated theatrics that are designed to boost bravado while also intimidating other states. Indeed, there is some substance to the publicized exercises, particularly in demonstrating the application of doctrinal concepts to certain scenarios. For instance, the Combined Maneuvers exercise in 2010 and other joint exercises displayed the DPRK's emphasis on integrated mobilization of the ground, sea, and air components for anti-access and amphibious operations. Nevertheless, much of the maneuvers propagated by the state media are only partial snapshots, making them insufficient in measuring the actual readiness of the KPA. Rather, the more critical training, exercises, and drills would be taking place behind closed doors.

The quality and effectiveness of training pivots on whether the forces train under realistic conditions and environment as well as improvisation to maximize their readiness for various contingencies. Yet despite the necessity, such training regimes raise cost issues. Although one could argue that the training regimes, military drills, and exercises are far less expensive to those of advanced states, the economic circumstances have nevertheless undermined the quality and quantity of training in the KPA. Particularly due to the chronic fuel shortages in the 1990s, the KPA has scaled back on large-scale live fire exercises and shifted to command post exercises, political education, and training that utilize simulators and rudimentary training aids.[125]

Among the various commands, the KPAAF in particular has borne the brunt of the fuel shortages, causing significant reduction in the flight hours for the pilots and operators both in aerial operations and in training. Referring to an unidentified report by the International Institute for Strategic Studies, Cordesman notes that the annual average flying hours for KPAAF pilots are approximately 20 hours which are much less than the pilots of the US forces who fly between 189 (combat) and 343 (airlift) hours.[126] Moreover, the US Department of the Army claims that much of the limited flight hours are devoted to takeoff and landing rather than combat maneuvers.[127] Such limitations have forced pilots to conduct onshore training using simulators and other training aids. While simulators and other equipment indeed have their share of merits, much is dependent on the quality of the systems used. Yet much of the simulators of the KPAAF are extremely rudimentary, with some being crudely retrofitted desktop computers that are limited in effectively simulating various conditions and maneuvers.

Changes are also seen in overseas education and training. In the early years, overseas education and training were vital in the more technical positions. For instance, the early generation of aviators received their training in foreign institutions such as the case of Ri Hwal and Wang Yon. Others, such as Jo Myong-rok graduated from both the Manchuria Aviation School and the Soviet Air Academy, and a number of other KPA elites such as O Kuk-ryol studied at the M.V. Frunze Military Academy. The earlier generations of KPA elites who were educated and trained overseas played key roles not only in command but also in formulating the operational concepts and doctrines. However, in the latter years of the Cold War era, there was a sharp decrease in the number of KPA personnel studying overseas, with the exception of some pilots who were sent to the USSR in the 1980s to undergo training to operate the MiG-23.[128] The decrease in the number of education and training abroad was due to not only the regional geopolitical circumstances but also the leadership's fears over possible factionalism that could undermine the regime. While the downturn in opportunities for study and training abroad was somewhat inevitable, it nevertheless shuns opportunities to modernize the KPA's operational and tactical concepts and doctrines.

In the past, the KPA personnel's aptitudes were occasionally supplemented (or compensated) through overseas operations. Since the armistice, the selected units and personnel were deployed overseas for military assistance. Although personnel were mostly sent as advisors, some engaged in actual combat such as in the Vietnam War and Yom Kippur War. Taking part in overseas deployments does allow the personnel to gain some real operational experience which could then be brought back and used to enhance the KPA's tactical doctrines and training regimes. Still, the actual numbers of personnel deployed to wars abroad were relatively modest and the deployments have been largely unpublicized. A significant part of this was due to the DPRK's limited capacity to finance overseas deployments and inadequate capabilities of the KPA. But much also had to do with Pyongyang's fears over the disruption to the force posture and that too much direct involvement (particularly in wars that involves the US) will bring the

DPRK into the firing line. Hence, given the limited size and nature of the KPA's overseas commitments, the actual benefits on the military's readiness have been limited.

Joint exercises with allies and likeminded states are also vital to enhance not only military readiness but also alliance relations. During the Cold War, the KPA took part in some joint exercises and exchanges with their Chinese and Soviet counterparts. The exercises were small-scale and irregular, often taking place when the Chinese and Soviet forces made visits to the DPRK and vice versa. Despite their low-profile nature, the interactions benefitted the DPRK. For instance, the joint exercises with the Soviet Pacific Fleet in July 1986 was a critical opportunity for the KPAN to learn not only about blue-water capabilities but also amphibious, naval mine, and submarine warfare.[129] However, the exercises were more for diplomatic reasons than building joint capabilities for a particular scenario, and the rationales to continue such initiatives waned as the Cold War came to a close. In the post-Cold War era, joint exercises have been minimal with reports claiming that China had refused to hold joint exercises with the DPRK.[130] Rather, China and Russia have conducted joint exercises such as Aerospace Security and Joint Sea in face of the recent tensions on the Korean peninsula – but without the DPRK. Indeed, it is entirely possible that North Korean officials participate in some Chinese and Russian exercises as observers, although the actual effects are far less than having the KPA being directly involved.

Given that the KPA has used the same platforms and equipment for a prolonged period of time, familiarity and expertise with their hardware would be high. Moreover, enlisted personnel are well-versed in their duties and missions as most of them stay in the same post throughout their long military service.[131] Taken together, one could argue that the DPRK could swiftly act when given the orders. However, actual readiness is not necessarily congruent with the time spent in the KPA. Particularly, since the dawn of the economic crisis, KPA personnel have been mobilized for civilian duties ranging from farming to construction and, in some cases, even dispatched to labor projects abroad. While such practices are based on the fact that the KPA is the pool of able-bodied persons fit for labor-intensive tasks, the over-utility of personnel for civilian duties comes at the expense of their military competence.

Problems are also evident in the readiness deficit between the active and reserve components. The RMTU serves as ready reserves that train alongside the regular KPA and are often assigned to combat roles while the WPRG and RYG are trained as local militias for civil defense, homeland security, and logistical roles.[132] Annual training hours differ greatly among the reserve units, where the RMTU undertakes approximately 500 hours per year, while the WPRG and RYG are said to undergo approximately 160 hours per year.[133] Moreover, both the WPRG and RYG are trained according to areas of residence, education institutions, or workplace, thus the nature of training and level of readiness would differ greatly – particularly since the economic downturn.[134] Of course, given that the large portion of the RMTU and WPRG are those that have previously served in the KPA, there is some level of readiness. Some discharged KPA personnel could be called back to serve their old roles – particularly for those with technical skills. The DPRK has conducted combined exercises and training of the active and reserve forces in 1975 and such programs are believed to have taken place on quasi-regular occasions.[135] Although one could argue that the training hours and readiness level of reservists are insignificant, problems will be inevitably felt during times of contingencies when the DPRK is compelled to rely on the reservists to supplement or compensate for losses in the KPA.

A key problem in the KPA's education and training regimes is the excessive level of ideological education. Although ideological education is inherent in any military institution, the intensity in the KPA is so much that it undermines the personnel's practical competence. One former KPA officer who later defected to South Korea claims that all soldiers are indoctrinated in "political thought" for two hours a day from Monday to Saturday.[136] Ideological education is more intense in the elite units where, for example, Bermudez claims that the elite 38th Airborne Brigade in the 1990s spent more than 25% of their training time on political education.[137] Even the internal KPA magazines and newspapers, such as Gunin Saenghwal and Joseon Inmingun talk little about concepts and technical matters relating to military readiness but are instead filled with articles, editorials, poems, songs, and stories about dedication, discipline, integrity, morale, and loyalty to the WPK leadership.[138]

Although the strong emphasis on ideological indoctrination is reflective of the regime's priority to ensure the KPA's political integrity and loyalty, the actual effects on military readiness are questionable. Indeed, the intense ideological indoctrination would be useful to promote camaraderie, discipline, and self-sacrifice among the KPA personnel to carry out their tasks even under the most challenging (actual or perceived) situations.[139] Still, ideological discipline cannot defy physiological limitations or completely alter human nature. History tells us that even the most ideologically disciplined soldier can be killed with a single bullet or a bayonet, become incapacitated by illnesses and shortage of food, or suffer from emotional stresses and illnesses. Moreover, in the case of the DPRK, the socio-economic issues are already affecting the KPA personnel, indicating that in the case of conflict, they will be mobilized in a physically, mentally, and morally fragile state.

The state's socio-economic issues also cause major problems for education and training regimes in the KPA. The deteriorating levels of health and morale combined with the problematic training regimes and equipment present a bad mix for the KPA. Although there is little doubt that the KPA is rigorously trained and familiarized with routines, this is largely because of the poor environment in which the personnel are forced to live in, and train with substandard equipment and training aids. In reality, overemphasis on political loyalty and severity of socio-economic issues affect the competence, health, morale, and professionalism of personnel that undermine the KPA's operational readiness. To overcome such problems, one method employed by the KPA is the frequent issuance of high alert status and mobilization for provocations and/or response to the US, ROK, and Japanese activities. Such an approach is not only useful in preventing decays in readiness as well as simulating war situations but also proving to the KPA that they must remain vigilant against the threats the nation faces. The downside, however, is that overdoing such politicized military orchestrations would cause fatigues, and the combination with the dire socio-economic circumstances is apt to corrode the personnel's faith in their duties and the regime.

Safety

Issues concerning safety practices in the KPA also warrant attention as problems would lead to injuries, deaths, damages, and malfunctions as well as deterioration in morale. While there is little available information on the accidents and incidents in North Korea, some reports have revealed the major safety problems in the KPA. For example, Yoo claims that in 1993, personnel in the Special Weapons Bureau were infected by mange due to the mishandling of biological agents.[140] Ballistic missiles are also not free from safety issues. On 28 April 2017, a Hwasong-12 IRBM launched from Pukchang Airfield in South Pyongan Province reportedly crashed into the city of Tokchon causing considerable damage.[141] Although the causes of the accidents and incidents are case-by-case, there are common, fundamental factors that lead to the safety problems in the KPA.

First, much of the safety problems in the KPA stem from the use of dated systems with lacking safety mechanics, making them far less fail safe than more modern systems. In particular, the DPRK's military modernization doctrine that involves the improvised and unconventional use of dated systems or those with makeshift upgrades mean that hardware is often used beyond their original specifications, increasing the likelihood of technical faults. Moreover, as the DPRK acquires more military assets, the KPA becomes responsible for implementing and practicing a greater variety of safety measures. Specifically, chemical, biological, and nuclear weapons are hard to maintain and operate due to their indiscriminate and sometimes unpredictable nature, raising questions over whether the DPRK can effectively control and manage such assets.

Second, while the DPRK regime emphasizes fluency in the use of equipment, the chronic shortages in energy and supplies have led to major compromises in training regimes, and the lack of adequate training aids raise questions concerning safety. Lacks in training not only corrode the personnel's familiarity with the equipment they use but also awareness and knowledge of response measures for the "what ifs" that are vital part of safe practices. Examples that illustrate the earlier are seen in the KPAAF, where the major compromises in flight hours combined with mechanical problems led to a surge in the number of crashes during the 1990s.[142]

Third, another key cause of the safety problems in the KPA is the hyper-centralized system. While centralized command and control may be effective in keeping the organization under control, there are questions over how the personnel at the lower echelons can autonomously and flexibly respond to problems at the field level. Although the DPRK has propagandistically emphasized safety in its tactical doctrines, the actual measures to enforce and oversee safety practices compared to those for security are unknown. Compounding the problem is the physical and psychological state of personnel, often worsened by the socio-economic circumstances in the Stalinist state. The various cases of misconduct reveal the major breaches of compliances and loopholes in the KPA, making it more vulnerable to deteriorating levels of safety awareness and practices.

For the KPA, fixing the issues in safety are difficult, requiring not only technological improvements but also structural and procedural reconfigurations from the top to the bottom echelons of the armed forces. Although the DPRK would be much concerned about the safety issues, it is dubious as to whether they are able to take the adequate steps to address the causes of the accidents and incidents.

Assessing the KPA's readiness

Although it is clear that the Line of Self-Reliant Defence shaped the KPA, Pyongyang has failed to find a balanced and coherent formula to attain and sustain an optimal level of military readiness. Above all, the chronic imbalance between force structural and operational readiness underscores Pyongyang's difficulties in effectively and efficiently mobilizing the KPA. The biggest contradiction is seen in how the KPA's force structure is built for massed, rapid warfare, while the operational readiness is not. Rather, the DPRK's mass armament with dated technologies upped the O&M costs beyond the state's capacity, consequently exacerbating the logistical deficiencies.

Even if the DPRK secures the resources and advance its military industries to import and produce indigenous variants of next-generation heavy-duty platforms, doing so would only take Pyongyang's defense planning dilemmas to a higher level. On the one hand, overhauling the KPA would implicate not only high switching costs in investment and O&M but also major reconfigurations in the military industry sector, operational and tactical concepts and doctrines, as well as education and training regimes. On the other hand, Pyongyang could increase the O&M provisions to the KPA and maximize the operational readiness of the existing units. However, considering the very size of the KPA, doing so requires massive amounts of resources and time to fix the array of outstanding problems, consequently undermining the capacity for acquisitions. While the DPRK is trying to find a balanced approach to advance both the force structural and operational readiness of the armed forces, the changes will not come into effect in the short term and could very well take years to yield any real results.

Indeed, the acquisition of strategic weapons was in large part a catch-all solution to the readiness imbalances in the KPA. From the defense planning viewpoint, Pyongyang viewed that ballistic missiles and nuclear weapons capabilities would be cost-effective means to enhance its firepower and power projection to fill the gaps in the KPA's force structural and operational readiness. Still, there are major risks in becoming over-dependent on WMD as they cannot be substitutes for conventional weapons. Despite the lethality of WMD, they do not necessarily dictate the outcome of war. In the Pacific War, the US's victory against Japan was not just because of the nuclear strikes on Hiroshima and Nagasaki but their ability to project their assets directly to the Japanese archipelago.

Furthermore, strategic weapons are of little use unless they are backed and covered by anti-access and area-denial capabilities. For the DPRK, the deficiencies in air and naval capabilities increase the vulnerability of the strategic weapons systems. Even if the DPRK constructed a series of fortified installations and fields an array of anti-ship and anti-air missiles, its anti-access and area-denial capabilities are nonetheless insufficient in filling the vulnerabilities against strikes by opposing forces. There are also issues concerning the tactical application of nuclear weapons to open up the defensive lines of the US and ROK forces. In particular, there are questions over whether the KPA has been adequately configured to operate under nuclear conditions such as not only the quality of armor to minimize the effects of radiation and EMP but also the ability to effectively execute operations while minimizing the disruptions caused by tactical nuclear attacks.

The imbalances in force structural and operational readiness also reveals how one of the most critical weaknesses of the KPA is the lack of advanced C4ISTAR systems to coordinate the various capabilities. The KPA units are disadvantaged not only on one-on-one terms but also more so in systems-vs-systems terms against technologically superior opponents such as the US, ROK, and Japan. Indeed, one could make the counterargument that the DPRK does not need high-end C4ISTAR systems and that such systems would even level the playing field in disadvantageous ways, exposing the KPA to cyber and electronic warfare attacks. However, such claims tend to overlook the nature of the KPA and the gravitational importance of C4ISTAR in modern warfare. In essence, C4ISTAR systems are vital not only in networking capabilities and creating a real-time common operational picture but also in effectively and efficiently coordinating logistics. Given the emphasis on working in multiple domains and fronts, as well as the assortment of both conventional and strategic weaponry, advanced C4ISTAR systems would play a critical roles for the KPA in executing coordinated operations and distributing the precious logistical assets to the units needed. Of course, the DPRK is far from being able to implement and operate the kind of C4ISTAR systems operated in advanced states. However, the DPRK could still assemble a partial C4ISTAR system that takes advantage of the fast developments in ICT to maximize the KPA's strengths and minimize its weaknesses.

The Line of Self-Reliant Defence has clearly shaped the North Korean armed forces with a mixture of capabilities purported for missions ranging from asymmetric operations to strategic strikes. The various weapons tests and military parades in recent years have revealed clear and consistent patterns of the DPRK's military modernization program. Still, there continues to be major deficiencies in the state's capacity to significantly modernize but more importantly provide adequate levels of operational readiness to sustain and operate the capabilities. The imbalances in the KPA's force structural and operational readiness inevitably undermine the DPRK's ability to achieve its strategic aims and objectives. Nonetheless, the KPA's readiness is still sufficient in inflicting enormous damage, and the slow but steady developments in the DPRK's military modernization in recent years evidence the threats that warrant greater attention.

Notes

1 International Institute for Strategic Studies, *The Military Balance 2020* (London, UK: International Institute for Strategic Studies, 2020), 284.
2 Ibid., 285.
3 Ibid.
4 Ibid.
5 Ibid.
6 Jeffrey Lewis, "More Rockets in Kim Jong Un's Pockets: North Korea Tests A New Artillery System," *38 North*, March 7, 2016, www.38north.org/2016/03/jlewis030716/; US Department of the Army, *ATP 7–100.2: North Korean Tactics* (Washington, DC: US Department of the Army, July 2020), A.3.
7 International Institute for Strategic Studies, *The Military Balance 2020*, 285.
8 Joseph S. Bermudez Jr., "P'okpoong: The KPA's New Main Battle Tank," *KPA Journal* 1, no. 4 (2010): 6.
9 Yong-won Yoo, "bukhan jaeraesik gunsaryeok mit mugichegye [North Korea's Conventional Military Force and Weapon Systems]," in *bukhangun sikeurit ripoteu [North Korea Military Secret Report]*, ed. Yong-won Yoo, Beom-chul Shin, and Jin-a Kim (Seoul, ROK: Planet Media, 2013), 188–89.
10 See: Joseph S. Bermudez Jr., "KPA Mechanized Infantry Battalion," *KPA Journal* 1, no. 7 (July 2010).
11 Yoo, "bukhan jaeraesik gunsaryeok mit mugichegye [North Korea's Conventional Military Force and Weapon Systems]," 211.
12 Ibid., 214.
13 International Institute for Strategic Studies, *The Military Balance 2020*, 285.
14 See: Jeffrey Lewis, "Oryx Blog on DPRK Arms Exports," *Arms Control Wonk*, June 25, 2014, www.armscontrolwonk.com/archive/207370/oryx-blog-on-dprk-arms-exports/.
15 See: Joseph S. Bermudez Jr., *North Korean Special Forces*, 2nd ed. (Annapolis, MD: Naval Institute Press, 1998); US Department of the Army, *ATP 7–100.2: North Korean Tactics.*
16 The Institute for Strategic Studies estimates 88,000, while the ROK Ministry for National Defense estimates 200,000. See: International Institute for Strategic Studies, *The Military Balance 2020*, 284; ROK Ministry of National Defense, *Defense White Paper 2018* (Seoul, ROK: ROK Ministry of National Defense, 2018), 39.
17 Soo-yeon Kim, "N.K. Sets Up Special Operation Forces amid Military Tensions," *Yonhap New Agency*, April 17, 2017, https://en.yna.co.kr/view/AEN20170417005352315.
18 Anthony H. Cordesman, *The Military Balance in the Koreas and Northeast Asia* (Washington, DC: Center for Strategic and International Studies, 2017), 115.
19 Yoo, "bukhan jaeraesik gunsaryeok mit mugichegye [North Korea's Conventional Military Force and Weapon Systems]," 316, 318.
20 Joseph S. Bermudez Jr., *The Armed Forces of North Korea* (St. Leonards, Australia: Allen & Unwin, 2001), 150–51; International Institute for Strategic Studies, *The Military Balance 2020*, 286.
21 International Institute for Strategic Studies, *The Military Balance 2020*, 285.
22 Joseph S. Bermudez Jr., "The Korean People's Navy Tests New Anti-ship Cruise Missile," *38 North*, February 8, 2015, www.38north.org/2015/02/jbermudez020815/.
23 International Institute for Strategic Studies, *The Military Balance 2020*, 285.
24 Ibid., 285.
25 Yoo, "bukhan jaeraesik gunsaryeok mit mugichegye [North Korea's Conventional Military Force and Weapon Systems]," 266.

26 Ankit Panda, "The Sinpo-C-Class: A New North Korean Ballistic Missile Submarine Is Under Construction," *The Diplomat*, October 18, 2017, https://thediplomat.com/2017/10/the-sinpo-c-class-a-new-north-korean-ballistic-missile-submarine-is-under-construction/; Joseph S. Bermudez Jr., "High Levels of Activity at North Korea's Sinpo South Shipyard," *38 North*, August 11, 2017, www.38north.org/2017/08/sinpo081117/; Joseph S. Bermudez Jr., "Sinpo South Shipyard: SLBM Test Not Imminent; Unknown Shipbuilding Program Underway," *38 North*, October 11, 2017, www.38north.org/2017/10/sinpo101117/; Joseph S. Bermudez Jr., "North Korea's Submarine Ballistic Missile Program Moves Ahead: Indications of Shipbuilding and Missile Ejection Testing," *38 North*, November 16, 2017, www.38north.org/2017/11/sinpo111617/.

27 See: Oliver Hotham, "New North Korean Submarine Capable of Carrying Three SLBMs: South Korean MND," *NK News*, July 31, 2019, www.nknews.org/2019/07/new-north-korean-submarine-capable-of-carrying-three-slbms-south-korean-mnd/; Joseph S. Bermudez Jr. and Victor D. Cha, "Sinpo South Shipyard: Construction of a New Ballistic Missile Submarine?" *Beyond Parallel*, August 28, 2019, https://beyondparallel.csis.org/sinpo-south-shipyard-construction-of-a-new-ballistic-missile-submarine/.

28 Rodong Sinmun, "uri sik sahoijueuigeonseoleul sae seungriero indohaneun widaehan tujaenggangryeong joseon rodongdang je8chadaehoieseo hasin gyeongaehaneun Kim Jong Un dongjieui bogoe daehayeo [Great Programme for Struggle Leading Korean-style Socialist Construction to Fresh Victory On Report Made by Supreme Leader Kim Jong Un at Eighth Congress of WPK]," *Rodong Sinmun*, January 9, 2021.

29 Yoo, "bukhan jaeraesik gunsaryeok mit mugichegye [North Korea's Conventional Military Force and Weapon Systems]," 253; Joseph S. Bermudez Jr., "New North Korean Helicopter Frigates Spotted," *38 North*, May 15, 2014, www.38north.org/2014/05/jbermudez051514/; Chosun Ilbo, "New Stripped Warship Spotted in N.Korean Port," *Chosun Ilbo*, November 7, 2007, http://english.chosun.com/site/data/html_dir/2007/11/07/2007110761022.html.

30 Bermudez Jr., "New North Korean Helicopter Frigates Spotted,"; Joost Oliemans and Stijn Mitzer, "A Navy Reborn: New Warships Spotted in North Korea," *NK Pro*, November 8, 2016, https://www.nknews.org/pro/a-navy-reborn-new-warships-spotted-in-north-korea/; Chad O'Carroll, "Exclusive: New Low-Visibility Corvette Spotted in North Korea," *NK News*, November 8, 2016, www.nknews.org/2016/11/exclusive-new-low-visibility-corvette-spotted-in-north-korea/.

31 Yoo, "bukhan jaeraesik gunsaryeok mit mugichegye [North Korea's Conventional Military Force and Weapon Systems]," 255.

32 International Institute for Strategic Studies, *The Military Balance 2020*, 285.

33 Ibid.

34 Ibid.

35 Ibid.; Hoon Jang, "bukhan haeguneui susangjeontuham [The North Korean Navy's Surface Combatants]," *Defense Today*, September 13, 2020, www.defensetoday.kr/news/articleView.html?idxno=1878.

36 H. I. Sutton, "North Korean VSV," *Covert Shores*, September 28, 2016, www.hisutton.com/North_Korean_VSV.html.

37 International Institute for Strategic Studies, *The Military Balance 2020*, 285; Yoo, "bukhan jaeraesik gunsaryeok mit mugichegye [North Korea's Conventional Military Force and Weapon Systems]," 272.

38 ROK Ministry of National Defense, *Defense White Paper 2018*, 30; Bermudez Jr., *The Armed Forces of North Korea*, 250; Joseph S. Bermudez Jr. and Beyond Parallel, "North Korean Special Operations Forces: Hovercraft Bases (Part I)," *Beyond Parallel*, January 25, 2018, https://beyondparallel.csis.org/north-korean-special-operations-forces-hovercraft-bases-part-1/; Joseph S. Bermudez Jr. and

Beyond Parallel, "North Korean Special Operations Forces: Hovercraft Bases (Part II)," *Beyond Parallel*, February 5, 2018, https://beyondparallel.csis.org/north-korean-special-operations-forces-hovercraft-bases-part-ii/; Joseph S. Bermudez Jr. and Beyond Parallel, "North Korean Special Operations Forces: Hovercraft Bases (Part III)," *Beyond Parallel*, February 15, 2018, https://beyondparallel.csis.org/north-korean-special-operations-forces-hovercraft-bases-part-iii/; Joseph S. Bermudez Jr. and Beyond Parallel, "North Korean Special Operations Forces: Hovercraft Bases (Part IV)," *Beyond Parallel*, March 5, 2018, https://beyondparallel.csis.org/north-korean-special-operations-forces-hovercraft-bases-part-iv/.

39 Some of these vessels belong to the RGB or the WPK rather than the KPAN.

40 Bermudez Jr., *The Armed Forces of North Korea*, 120.

41 International Institute for Strategic Studies, *The Military Balance 2020*, 285.

42 See: Joseph S. Bermudez Jr., "Korean People's Navy 14.5mm 6 Barrel CIWS," *KPA Journal* 2, no. 8 (August 2012); Joseph S. Bermudez Jr., "Korean People's Navy 30mm CIWS," *KPA Journal* 2, no. 12 (December 2013): 1.

43 Bermudez Jr., *The Armed Forces of North Korea*, 121; International Institute for Strategic Studies, *The Military Balance 2020*, 285.

44 Bermudez Jr., *The Armed Forces of North Korea*, 121.

45 International Institute for Strategic Studies, *The Military Balance 2020*, 286.

46 For examples, see: Curtis Melvin and Joseph S. Bermudez Jr, "KPA Navy Upgrades in the East Sea," *38 North*, September 1, 2016, www.38north.org/2016/09/munchon090116/.

47 International Institute for Strategic Studies, *The Military Balance 2020*, 286; US Department of the Army, *ATP 7-100.2: North Korean Tactics*, B.1.

48 International Institute for Strategic Studies, *The Military Balance 2020*, 286.

49 Ibid., 286; Bermudez Jr., *The Armed Forces of North Korea*, 147–48.

50 International Institute for Strategic Studies, *The Military Balance 2020*, 286.

51 See: Joseph S. Bermudez Jr., "MiG-29 in KPAF Service," *KPA Journal* 2, no. 4 (April 2011): 6–11.

52 International Institute for Strategic Studies, *The Military Balance 2020*, 286.

53 Bermudez Jr., *The Armed Forces of North Korea*, 149.

54 International Institute for Strategic Studies, *The Military Balance 2020*, 286.

55 Ibid., 286.

56 Bermudez Jr., *The Armed Forces of North Korea*, 151.

57 International Institute for Strategic Studies, *The Military Balance 2020*, 286.

58 US Department of the Army, *ATP 7-100.2: North Korean Tactics*, D.3; Yoo, "bukhan jaeraesik gunsaryeok mit mugichegye [North Korea's Conventional Military Force and Weapon Systems]," 332.

59 International Institute for Strategic Studies, *The Military Balance 2020*, 286.

60 Ibid., 285.

61 US Department of the Army, *ATP 7-100.2: North Korean Tactics*, D.3.

62 International Institute for Strategic Studies, *The Military Balance 2020*, 286.

63 Ibid., 286.

64 See: Bermudez Jr., *The Armed Forces of North Korea*, 149–52.

65 Ibid., 150; Reports also claimed that North Korea's An-2 is capable of mounting KN-01 missiles. Kwi-geun Kim, "buk, seohae sanggong AN-2giseo misail 2 bal balsa [North Korea – AN-2 Fires Two Missiles in the West Sea]," *Yonhap News*, October 9, 2008, www.yonhapnews.co.kr/bulletin/2008/10/09/0200000000AKR20081009046500043.HTML.

66 Yoo, "bukhan jaeraesik gunsaryeok mit mugichegye [North Korea's Conventional Military Force and Weapon Systems]," 306, 309.

67 Joseph S. Bermudez Jr., "North Korea Drones On: Redeux," *38 North*, January 19, 2016, www.38north.org/2016/01/jbermudez011916/.

68 Cheol-hwan Kang and Yong-hyun An, "Kim Jong-il: jungguke choisinye jeontugi jiwon yocheong [Kim Jong-il: Requests Provision of Modern Fighter Jets from China]," *Chosun Ilbo*, June 17, 2010, http://nk.chosun.com/news/news.html?ACT=detail&res_id=126058; Zachary Keck, "North Korea Wants to Buy Russia's Super Advanced Su-35 Fighter Jet," *The National Interest*, January 9, 2015, http://nationalinterest.org/blog/the-buzz/north-korea-wants-buy-russias-super-advanced-su-35-fighter-12005.
69 Frank V. Pabian, Joseph S. Bermudez Jr, and Jack Liu, "North Korea's Punggye-ri Nuclear Test Site: Satellite Imagery Shows Post-Test Effects and New Activity in Alternate Tunnel Portal Areas," *38 North*, September 12, 2017, 38north.org/2017/09/punggye091217/.
70 Siegfried S. Hecker, "What I Found in North Korea," *Foreign Affairs*, December 9, 2010, www.foreignaffairs.com/articles/northeast-asia/2010-12-09/what-i-found-north-korea.
71 See: 38 North, "North Korea's Yongbyon Nuclear Research Center: Testing of Reactor Cooling Systems; Construction of Two New Non-Industrial Buildings," July 6, 2018, www.38north.org/2018/07/yongbyon070618/.
72 Ji-eun Kim and Sung-hui Moo, "North Korea Sets Up Special Force for Radioactive Bomb Attacks," *Radio Free Asia*, August 29, 2016, www.rfa.org/english/news/korea/north-korea-sets-up-special-force-for-radioactive-bomb-attacks-08262016162017.html.
73 ROK Ministry of National Defense, *Defense White Paper 2018*, 34.
74 International Crisis Group, "North Korea's Chemical and Biological Weapons Programs," in *Asia Report* (Seoul, ROK and Brussels, Belgium: International Crisis Group, 2009), 6.
75 Ibid.
76 Yong-won Yoo, "bukhan bidaeching jeonryeok [North Korea's Assymetric Military Capabilities]," in *bukhangun sikeurit ripoteu [North Korea Military Secret Report]*, ed. Yong-won Yoo, Beom-chul Shin, and Jin-a Kim (Seoul, ROK: Planet Media, 2013), 157–59; Hyun-kyung Kim, Elizabeth Philipp, and Hattie Chung, "The Known and Unknown: North Korea's Biological Weapons Program," in *Project on Managing the Microbe* (Cambridge, MA: Belfer Center for Science and International Affairs, 2017), 5.
77 Missile Defense Project, "KN-08/Hwasong 13," *Missile Threat*, August 8, 2016, https://missilethreat.csis.org/missile/kn-08/; Missile Defense Project, "Hwasong-14 (KN-20)," *Missile Threat*, July 27, 2017, https://missilethreat.csis.org/missile/hwasong-14/; Missile Defense Project, "Hwasong-15 (KN-22)," *Missile Threat*, December 7, 2017, https://missilethreat.csis.org/missile/hwasong-15-kn-22/; Missile Defense Project, "KN-14 (KN-08 Mod 2)," *Missile Threat*, August 8, 2016, https://missilethreat.csis.org/missile/kn-14/.
78 Missile Defense Project, "Hwasong-12," *Missile Threat*, May 16, 2017, https://missilethreat.csis.org/missile/hwasong-12/.
79 Missile Defense Project, "No Dong 1," *Missile Threat*, August 9, 2016, https://missilethreat.csis.org/missile/no-dong/; Missile Defense Project, "Hwasong-9 (Scud-ER)," *Missile Threat*, August 8, 2016, https://missilethreat.csis.org/missile/scud-er/.
80 Missile Defense Project, "Pukguksong-2 (KN-15)," *Missile Threat*, March 5, 2017, https://missilethreat.csis.org/missile/pukkuksong-2/.
81 Missile Defense Project, "Hwasong-5 ('Scud B' Variant)," *Missile Threat*, August 8, 2016, https://missilethreat.csis.org/missile/hwasong-5/; Missile Defense Project, "Hwasong-6 ('Scud C' Variant)," *Missile Threat*, August 8, 2016, https://missilethreat.csis.org/missile/hwasong-6/.
82 Missile Defense Project, "KN-23," *Missile Threat*, July 1, 2019, https://missilethreat.csis.org/missile/kn-23/; Missile Defense Project, "KN-24," *Missile*

Threat, June 23, 2020, https://missilethreat.csis.org/missile/kn-24/; Missile Defense Project, "KN-25," *Missile Threat*, June 23, 2020, https://missilethreat.csis.org/missile/kn-25/.

83 Missile Defense Project, "KN-18 (Scud MaRV Variant)," *Missile Threat*, April 18, 2017, https://missilethreat.csis.org/missile/kn-18-marv-scud-variant/.

84 Jeffrey Lewis, "DPRK SLBM Test," *Arms Control Wonk*, May 13, 2015, www.armscontrolwonk.com/archive/207631/dprk-slbm-test/. North Korea also reportedly built another test barge at Nampho Naval Base. See: Joseph S. Bermudez Jr., "North Korea's Submarine-Launched Ballistic Missile Program Advances: Second Missile Test Stand Barge Almost Operational," *38 North*, December 1, 2017, www.38north.org/2017/12/nampo120117/.

85 Missile Defense Project, "Pukguksong-3 (KN-26)," *Missile Threat*, October 7, 2019, https://missilethreat.csis.org/missile/pukguksong-3/.

86 For China, the R&D and deployment of the Julang-1 took almost two decades, and Russia who has long experience has encountered problems with its latest RSM-56 (SS-NX-30 or SS-N-32) missiles.

87 See: Joseph S. Bermudez Jr., "The KN-02 SRBM," *KPA Journal* 1, no. 2 (February 2010); Joseph S. Bermudez Jr., "The Scud B SRBM in KPA Service," *KPA Journal* 1, no. 3 (March 2010).

88 US Senate Committee on Governmental Affairs Subcommittee on International Security, Proliferation and Federal Services, *Senate Hearing 105–241: North Korean Missile Proliferation*, October 21, 1997, 14.

89 See: Jeffrey Lewis, "North Korea's Nuclear Weapons: The Great Miniaturization Debate," *38 North*, February 5, 2015, www.38north.org/2015/02/jlewis020515/.

90 See: Michael Elleman, "Video Casts Doubt on North Korea's Ability to Field an ICBM Re-entry Vehicle," July 31, 2017, www.38north.org/2017/07/melleman073117/.

91 Rodong Sinmun, "uri sik sahoijueuigeonseoleul sae seungriero indohaneun widaehan tujaenggangryeong joseon rodongdang je8chadaehoieseo hasin gyeongaehaneun Kim Jong Un dongjieui bogoe daehayeo [Great Programme for Struggle Leading Korean-style Socialist Construction to Fresh Victory On Report Made by Supreme Leader Kim Jong Un at Eighth Congress of WPK]," *Rodong Sinmun*, January 9, 2021.

92 For deeper analysis on the DPRK's cyber warfare program, see: Jenny Jun, LaFoy Scott, Ethan Sohn, James Andrew Lewis, and Victor D. Cha, *North Korea's Cyber Operations: Strategy and Responses* (Washington, DC and Lanham, MD: Center for Strategic and International Studies and Rowman & Littlefield, December 2015); Michael Raska, "North Korea's Evolving Cyber Strategies: Continuity and Change," *SIRIUS* 4, no. 2 (2020).

93 See: So-yeol Kim, "Defense Systems Hacking on the Increase," *Daily NK*, June 17, 2009, www.dailynk.com/english/read.php?cataId=nk00100&num=5058.

94 James A. Lewis, "North Korea and Cyber Catastrophe – Don't Hold Your Breath," *38 North*, January 12, 2018, www.38north.org/2018/01/jalewis011218/.

95 Joseph S. Bermudez Jr., "SIGINT, EW, and EIW in the Korean People's Army: An Overview of Development and Organization," in *Bytes and Bullets in Korea*, ed. Alexandre Y. Mansourov (Honolulu, HI: APCSS, 2005), 244.

96 Ibid., 250–51, 269.

97 Bermudez Jr., *The Armed Forces of North Korea*, 115, 151; Bermudez Jr., "SIGINT, EW, and EIW in the Korean People's Army: An Overview of Development and Organization," 252–74.

98 Chosun Ilbo, "GPS Jamming from N.Korea Hit 1,137 S.Korean Aircraft," *Chosun Ilbo*, September 19, 2012, http://english.chosun.com/site/data/html_dir/2012/09/19/2012091900628.html.

99 Steve Herman, "Secret Manual Gives Glimpse of North Korean Military Tactics," *Voice of America*, September 18, 2010, www.voanews.com/a/secret-manual-gives-glimpse-of-north-korean-military-tactics-103253534/126266.html.
100 Bermudez Jr., "SIGINT, EW, and EIW in the Korean People's Army: An Overview of Development and Organization," 256, 260–61.
101 Andrew Scobell and John M. Sanford, *North Korea's Military Threat: Pyongyang's Conventional Forces, Weapons of Mass Destruction, and Ballistic Missiles* (Carlisle, PA: US Army War College Strategic Studies Institute, 2007), 56.
102 Yoo, "bukhan jaeraesik gunsaryeok mit mugichegye [North Korea's Conventional Military Force and Weapon Systems]," 306; Bermudez Jr., "MiG-29 in KPAF Service," 2.
103 US Department of the Army, *ATP 7–100.2: North Korean Tactics*, E.3.
104 Joseph S. Bermudez Jr., "KPA Land-based MR-104 DRUM TILT Radar," *KPA Journal* 2, no. 9 (September 2012); US Department of the Army, *ATP 7–100.2: North Korean Tactics*, E.3.
105 US Department of the Army, *ATP 7–100.2: North Korean Tactics*, E.3.
106 Bermudez Jr., "SIGINT, EW, and EIW in the Korean People's Army: An Overview of Development and Organization," 244–245.
107 Ibid., 256, 259.
108 For more on the C4ISTAR systems sold by Glocom, see: James Bingham, "Details of North Korean C4I Systems Emerge," *IHS Jane's International Defence Review*, March 16, 2017, www.janes.com/article/68771/details-of-north-korean-c4i-systems-emerge; Andrea Berger, "GLOCOM is at it Again," *Arms Control Wonk*, August 22, 2017, www.armscontrolwonk.com/archive/1203749/glocom-is-at-it-again/.
109 Peter J. Brown, "Is North Korea Using China's Satellites to Guide its Missiles?" *Asia Times Online*, May 24, 2017, www.atimes.com/article/north-korea-using-chinas-satellites-guide-missiles/.
110 Cordesman, *The Military Balance in the Koreas and Northeast Asia*, 152.
111 Jiro Ishimaru, "gasorin kyuutoushi saikouchikiroku gunmo mokutansyato gyuusyade bussiunpan seisai eikyou jiwari [Effects of Economic Sanctions: Gasoline Prices Exponentially Increases to Record Levels – Even the Military is Using Wood-Combusted Vehicles and Bullock Carts to Transport Goods]," *Asia Press International*, January 8, 2018, www.asiapress.org/apn/author-list/ishimaru-jiro/post-57261/.
112 See: Jeffrey Lewis, "Domestic UDMH Production in the DPRK," *Arms Control Wonk*, September 27, 2017, www.armscontrolwonk.com/archive/1204170/domestic-udmh-production-in-the-dprk/.
113 See: Ryo Hinata-Yamaguchi, "North Korea's Transport Policies: Current Status and Problems," *Journal of International Relations* 19, no. 2 (2016).
114 Andrei N. Lankov, *North of the DMZ: Essays on Daily Life in North Korea* (Jefferson, NC: McFarland & Company, 2007), 156–58; Japan External Trade Organization, "2014nendo saikinno kitachousenkeizaini kansuru chousa [Assessment of the North Korean Economy for the FY2014]," (Tokyo, Japan: Japan External Trade Organization, 2015), 104.
115 Byung-min Ahn, *gyogwaseoe annaoneun bukhaneui gyotong iyagi [North Korea's Transport not Discussed in Textbooks]* (Seoul, ROK: Institution for Unification Education, 2014), 54.
116 See: Joseph S. Bermudez Jr., "KPA Engineer River Crossing Forces," *KPA Journal* 1, no. 8 (August 2010); Joseph S. Bermudez Jr., "Wartime 'Underwater' Bridges," *KPA Journal* 2, no. 1 (January 2011).
117 For instance, Russia stopped providing parts for the MiG-29 to the DPRK. See: Sung-bin Choi, Jae-moon Yoo, and Si-woo Kwak, *bukhan gunsusaneob*

gaehwang [Current State of the North Korean Military Industry] (Seoul, ROK: Korea Institute for Defense Analyses, 2005), 32.

118 Bermudez Jr., *The Armed Forces of North Korea*, 143.

119 International Institute for Strategic Studies, *The Military Balance 2020*, 286.

120 Ibid., 286; ROK Ministry of National Defense, *Defense White Paper 2018*, 35.

121 Bermudez Jr., *The Armed Forces of North Korea*, 82.

122 See: Seok-young Lee, "Officer Families Just Like the Rest," *Daily NK*, December 2, 2011, www.dailynk.com/english/read.php?cataId=nk01500&num=8475.

123 Baekgwasajeon Chulpansa, *joseon daebaekgwasajeon [Encyclopedia of Korea]*, vol. 3 (Pyongyang, DPRK: Baekgwasajeon Chulpansa, 1995), 272.

124 Bermudez Jr., *The Armed Forces of North Korea*, 104.

125 For examples of North Korean training aids, see: Joseph S. Bermudez Jr., "KPA Tank Training Aids," *KPA Journal* 2, no. 10 (October 2012); Joseph S. Bermudez Jr., "KPN Training Aid," *KPA Journal* 2, no. 12 (December 2013); Cordesman, *The Military Balance in the Koreas and Northeast Asia*, 149.

126 Cordesman, "The Military Balance in the Koreas and Northeast Asia," 148–49. Some reports in 2012 claimed that the KPAAF significantly increased the flight training sorties, although still far short of the counterparts in the US, ROK, and Japan. See: Chosun Ilbo, "N.Korea Steps Up Air Force Training Flights," *Chosun Ilbo*, May 29, 2012, http://english.chosun.com/site/data/html_dir/2012/03/29/2012032901309.html.

127 US Department of the Army, *ATP 7-100.2: North Korean Tactics*, B.2.

128 Katsuichi Tsukamoto, *kitachousengunto seiji [The North Korean Army and Politics]* (Tokyo, Japan: Hara Shobo, 2000), 140.

129 Bermudez Jr., *The Armed Forces of North Korea*, 105.

130 So-yeol Kim, "China Ducks Joint Exercises Proposal," *Daily NK*, August 8, 2011, www.dailynk.com/english/read.php?cataId=nk00100&num=8033.www.dailynk.com/english/read.php?cataId=nk00100&num=8033.

131 Scobell and Sanford, *North Korea's Military Threat: Pyongyang's Conventional Forces, Weapons of Mass Destruction, and Ballistic Missiles*, 67.

132 For instance, Scalapino and Lee argue that the WPRG is "[S]o organized that its various essential functions are allocated among those having the necessary skills or responsibilities, who are already serving at the level involved." Robert A. Scalapino and Chong-sik Lee, *Communism in Korea* (Berkeley, CA: University of California Press, 1972), 948.

133 ROK Ministry of Unification, *Understanding North Korea* (Seoul, ROK: ROK Ministry of Unification, 2017), 155–56.

134 For details on the organization of the WPRG and RYG, see: Bermudez Jr., *The Armed Forces of North Korea*, 162–69.

135 Yang-ju Kwon, *bukhangunsaeui ihae [The Comprehension of North Korean Military]*, Expanded ed. (Seoul, ROK: Korea Institute of Defense Analyses, 2014), 166.

136 Jung-yeon Lee, *bukhanguneneun geonbbangi eobda [No Dry Biscuits in the North Korean Army]* (Seoul, ROK: Flash Media, 2007), 151.

137 Bermudez Jr., *North Korean Special Forces*, 235.

138 See: Joseon Inmingun Chulpansa, "eoryeoun immureul naege [Give the Difficult Missions to Me]," *gunin saenghwal [Lives of Soldiers]* 2 (2008): 41; Joseon Inmingun Chulpansa, "jeonsadeuleul wihayeo jihwigwani itda [Commanders are there for the Warriors]," *gunin saenghwal [Lives of Soldiers]* 2 (2008): 20; Joseon Inmingun Chulpansa, "hunryeonseonggwawa jihwigwaneui yoguseong [Training Performance and Requirements of Commanders]," *gunin saenghwal [Lives of Soldiers]* 9 (2008): 49–51.

139 For more elaborate discussions on nationalism and their effect on militaries, see: Barry R. Posen, "Nationalism, the Mass Army, and Military Power," *International Security* 18, no. 2 (1993): 81.
140 Yoo, "bukhan bidaeching jeonryeok [North Korea's Assymetric Military Capabilities]," 157.
141 Ankit Panda and Dave Schmerler, "When a North Korean Missile Accidentally Hit a North Korean City," *The Diplomat*, January 3, 2018, https://thediplomat.com/2018/01/when-a-north-korean-missile-accidentally-hit-a-north-korean-city/.
142 Bermudez Jr., *The Armed Forces of North Korea*, 146–47.

References

38 North. "North Korea's Yongbyon Nuclear Research Center: Testing of Reactor Cooling Systems; Construction of Two New Non-Industrial Buildings." July 6, 2018. www.38north.org/2018/07/yongbyon070618/.

Ahn, Byung-min. *gyogwaseoe annaoneun bukhaneui gyotong iyagi [North Korea's Transport not Discussed in Textbooks]*. Seoul, ROK: Institution for Unification Education, 2014.

Baekgwasajeon Chulpansa. *joseon daebaekgwasajeon [Encyclopedia of Korea]*. Vol. 3. Pyongyang, DPRK: Baekgwasajeonchulpansa, 1995.

Berger, Andrea. "GLOCOM is at it Again." *Arms Control Wonk*, August 22, 2017. www.armscontrolwonk.com/archive/1203749/glocom-is-at-it-again/.

Bermudez Jr., Joseph S. *North Korean Special Forces*. 2nd ed. Annapolis, MD: Naval Institute Press, 1998.

———. *The Armed Forces of North Korea*. St. Leonards, Australia: Allen & Unwin, 2001.

———. "SIGINT, EW, and EIW in the Korean People's Army: An Overview of Development and Organization." In *Bytes and Bullets in Korea*, edited by Alexandre Y. Mansourov. Honolulu, HI: APCSS, 2005.

———. "The KN-02 SRBM." *KPA Journal* 1, no. 2 (February 2010).

———. "KPA Engineer River Crossing Forces." *KPA Journal* 1, no. 8 (August 2010).

———. "KPA Mechanized Infantry Battalion." *KPA Journal* 1, no. 7 (July 2010).

———. "P'okpoong: The KPA's New Main Battle Tank." *KPA Journal* 1, no. 4 (2010).

———. "The Scud B SRBM in KPA Service." *KPA Journal* 1, no. 3 (March 2010).

———. "MiG-29 in KPAF Service." *KPA Journal* 2, no. 4 (April 2011).

———. "Wartime 'Underwater' Bridges." *KPA Journal* 2, no. 1 (January 2011).

———. "Korean People's Navy 14.5mm 6 Barrel CIWS." *KPA Journal* 2, no. 8 (August 2012).

———. "KPA Land-based MR-104 DRUM TILT Radar." *KPA Journal* 2, no. 9 (September 2012).

———. "KPA Tank Training Aids." *KPA Journal* 2, no. 10 (October 2012).

———. "Korean People's Navy 30mm CIWS." *KPA Journal* 2, no. 12 (December 2013).

———. "KPN Training Aid." *KPA Journal* 2, no. 12 (December 2013).

———. "New North Korean Helicopter Frigates Spotted." *38 North*, May 15, 2014. www.38north.org/2014/05/jbermudez051514/.

———. "The Korean People's Navy Tests New Anti-Ship Cruise Missile." *38 North*, February 8, 2015. www.38north.org/2015/02/jbermudez020815/.
———. "North Korea Drones On: Redeux." *38 North*, January 19, 2016. www.38north.org/2016/01/jbermudez011916/.
———. "High Levels of Activity at North Korea's Sinpo South Shipyard." *38 North*, August 11, 2017. www.38north.org/2017/08/sinpo081117/.
———. "North Korea's Submarine Ballistic Missile Program Moves Ahead: Indications of Shipbuilding and Missile Ejection Testing." *38 North*, November 16, 2017. www.38north.org/2017/11/sinpo111617/.
———. "North Korea's Submarine-Launched Ballistic Missile Program Advances: Second Missile Test Stand Barge Almost Operational." *38 North*, December 1, 2017. www.38north.org/2017/12/nampo120117/.
———. "Sinpo South Shipyard: SLBM Test Not Imminent; Unknown Shipbuilding Program Underway." *38 North*, October 11, 2017. www.38north.org/2017/10/sinpo101117/.
Bermudez Jr., Joseph S., and Victor D. Cha. "Sinpo South Shipyard: Construction of a New Ballistic Missile Submarine?" *Beyond Parallel*, August 28, 2019. https://beyondparallel.csis.org/sinpo-south-shipyard-construction-of-a-new-ballistic-missile-submarine/.
Bermudez Jr., Joseph S., and Beyond Parallel. "North Korean Special Operations Forces: Hovercraft Bases (Part I)." *Beyond Parallel*, January 25, 2018. https://beyondparallel.csis.org/north-korean-special-operations-forces-hovercraft-bases-part-1/.
———. "North Korean Special Operations Forces: Hovercraft Bases (Part II)." *Beyond Parallel*, February 5, 2018. https://beyondparallel.csis.org/north-korean-special-operations-forces-hovercraft-bases-part-ii/.
———. "North Korean Special Operations Forces: Hovercraft Bases (Part III)." *Beyond Parallel*, February 15, 2018. https://beyondparallel.csis.org/north-korean-special-operations-forces-hovercraft-bases-part-iii/.
———. "North Korean Special Operations Forces: Hovercraft Bases (Part IV)." *Beyond Parallel*, March 5, 2018. https://beyondparallel.csis.org/north-korean-special-operations-forces-hovercraft-bases-part-iv/.
Bingham, James. "Details of North Korean C4I Systems Emerge." *IHS Jane's International Defence Review*, March 16, 2017. www.janes.com/article/68771/details-of-north-korean-c4i-systems-emerge.
Brown, Peter J. "Is North Korea Using China's Satellites to Guide its Missiles?" *Asia Times Online*, May 24, 2017. www.atimes.com/article/north-korea-using-chinas-satellites-guide-missiles/.
Choi, Sung-bin, Jae-moon Yoo, and Si-woo Kwak. *bukhan gunsusaneob gaehwang [Current State of the North Korean Military Industry]*. Seoul, ROK: Korea Institute for Defense Analyses, 2005.
Chosun Ilbo. "New Stripped Warship Spotted in N.Korean Port." *Chosun Ilbo*, November 7, 2007. http://english.chosun.com/site/data/html_dir/2007/11/07/2007110761022.html.
———. "N.Korea Steps Up Air Force Training Flights." *Chosun Ilbo*, May 29, 2012. http://english.chosun.com/site/data/html_dir/2012/03/29/2012032901309.html.
———. "GPS Jamming from N.Korea Hit 1,137 S.Korean Aircraft." *Chosun Ilbo*, September 19, 2012. http://english.chosun.com/site/data/html_dir/2012/09/19/2012091900628.html.
Cordesman, Anthony H. *The Military Balance in the Koreas and Northeast Asia*. Washington, DC: Center for Strategic and International Studies, 2017.

Elleman, Michael. "Video Casts Doubt on North Korea's Ability to Field an ICBM Re-entry Vehicle." *38 North*, July 31, 2017. www.38north.org/2017/07/melleman073117/.

Hecker, Siegfried S. "What I Found in North Korea." *Foreign Affairs*, December 9, 2010. www.foreignaffairs.com/articles/northeast-asia/2010-12-09/what-i-found-north-korea.

Herman, Steve. "Secret Manual Gives Glimpse of North Korean Military Tactics." *Voice of America*, September 18, 2010. www.voanews.com/a/secret-manual-gives-glimpse-of-north-korean-military-tactics-103253534/126266.html.

Hinata-Yamaguchi, Ryo. "North Korea's Transport Policies: Current Status and Problems." *Journal of International Relations* 19, no. 2 (2016).

Hotham, Oliver. "New North Korean Submarine Capable of Carrying Three SLBMs: South Korean MND." *NK News*, July 31, 2019. www.nknews.org/2019/07/new-north-korean-submarine-capable-of-carrying-three-slbms-south-korean-mnd/.

International Crisis Group. "North Korea's Chemical and Biological Weapons Programs." In *Asia Report*. Seoul, ROK and Brussels, Belgium: International Crisis Group, 2009.

International Institute for Strategic Studies. *The Military Balance 2020*. London, UK: International Institute for Strategic Studies, 2020.

Ishimaru, Jiro. "gasorin kyuutoushi saikouchikiroku gunmo mokutansyato gyuusyade bussiunpan seisai eikyou jiwari [Effects of Economic Sanctions: Gasoline Prices Exponentially Increases to Record Levels – Even the Military is Using Wood-Com busted Vehicles and Bullock Carts to Transport Goods]." *Asia Press International*, January 8, 2018. www.asiapress.org/apn/author-list/ishimaru-jiro/post-57261/.

Jang, Hoon. "bukhan haeguneui susangjeontuham [The North Korean Navy's Surface Combatants]." *Defense Today*, September 13, 2020. www.defensetoday.kr/news/articleView.html?idxno=1878.

Japan External Trade Organization. *2014nendo saikinno kitachousenkeizaini kansuru chousa [Assessment of the North Korean Economy for the FY2014]*. (Tokyo, Japan: Japan External Trade Organization, 2015).

Joseon Inmingun Chulpansa. "eoryeoun immureul naege [Give the Difficult Missions to Me]." *gunin saenghwal [Lives of Soldiers]* 2 (2008).

———. "hunryeonseonggwawa jihwigwaneui yoguseong [Training Performance and Requirements of Commanders]." *gunin saenghwal [Lives of Soldiers]* 9 (2008).

———. "jeonsadeuleul wihayeo jihwigwani itda [Commanders are there for the Warriors]." *gunin saenghwal [Lives of Soldiers]* 2 (2008).

Jun, Jenny, LaFoy Scott, Ethan Sohn, James Andrew Lewis, and Victor D. Cha. *North Korea's Cyber Operations: Strategy and Responses*. Washington, DC and Lanham, MD: Center for Strategic and International Studies and Rowman & Littlefield, December 2015.

Kang, Cheol-hwan, and Yong-hyun An. "Kim Jong-il: jungguke choisinye jeontugi jiwon yocheong [Kim Jong-il: Requests Provision of Modern Fighter Jets from China]." *Chosun Ilbo*, June 17, 2010. http://nk.chosun.com/news/news.html?ACT=detail&res_id=126058.

Keck, Zachary. "North Korea Wants to Buy Russia's Super Advanced Su-35 Fighter Jet." *The National Interest*, January 9, 2015. http://nationalinterest.org/blog/the-buzz/north-korea-wants-buy-russias-super-advanced-su-35-fighter-12005.

Kim, Hyun-kyung, Elizabeth Philipp, and Hattie Chung. "The Known and Unknown: North Korea's Biological Weapons Program." In *Project on Managing the Microbe*. Cambridge, MA: Belfer Center for Science and International Affairs, 2017.

Kim, Ji-eun, and Sung-hui Moo. "North Korea Sets Up Special Force for Radioactive Bomb Attacks." *Radio Free Asia*, August 29, 2016. www.rfa.org/english/news/korea/north-korea-sets-up-special-force-for-radioactive-bomb-attacks-08262016162017.html.

Kim, Kwi-geun. "buk, seohae sanggong AN-2giseo misail 2 bal balsa [North Korea – AN-2 Fires Two Missiles in the West Sea]." *Yonhap News*, October 9, 2008. www.yonhapnews.co.kr/bulletin/2008/10/09/0200000000AKR20081009046500043.HTML.

Kim, So-yeol. "Defense Systems Hacking on the Increase." *Daily NK*, June 17, 2009. www.dailynk.com/english/read.php?cataId=nk00100&num=5058.

———. "China Ducks Joint Exercises Proposal." *Daily NK*, August 8, 2011. www.dailynk.com/english/read.php?cataId=nk00100&num=8033.

Kim, Soo-yeon. "N.K. Sets Up Special Operation Forces amid Military Tensions." *Yonhap New Agency*, 17 April 2017. https://en.yna.co.kr/view/AEN20170417005352315.

Kwon, Yang-ju. *bukhangunsaeui ihae [The Comprehension of North Korean Military]*. Expanded ed. Seoul, ROK: Korea Institute of Defense Analyses, 2014.

Lankov, Andrei N. *North of the DMZ: Essays on Daily Life in North Korea*. Jefferson, NC: McFarland & Company, 2007.

Lee, Jung-yeon. *bukhanguneneun geonbbangi eobda [No Dry Biscuits in the North Korean Army]*. Seoul, ROK: Flash Media, 2007.

Lee, Seok-young. "Officer Families Just Like the Rest." *Daily NK*, December 2, 2011. www.dailynk.com/english/read.php?cataId=nk01500&num=8475.

Lewis, James A. "North Korea and Cyber Catastrophe – Don't Hold Your Breath." *38 North*, January 12, 2018. www.38north.org/2018/01/jalewis011218/.

Lewis, Jeffrey. "Oryx Blog on DPRK Arms Exports." *Arms Control Wonk*, June 25, 2014. www.armscontrolwonk.com/archive/207370/oryx-blog-on-dprk-arms-exports/.

———. "North Korea's Nuclear Weapons: The Great Miniaturization Debate." *38 North*, February 5, 2015. www.38north.org/2015/02/jlewis020515/.

———. "DPRK SLBM Test," *Arms Control Wonk*, May 13, 2015. www.armscontrolwonk.com/archive/207631/dprk-slbm-test/.

———. "More Rockets in Kim Jong Un's Pockets: North Korea Tests A New Artillery System." *38 North*, March 7, 2016. www.38north.org/2016/03/jlewis030716/.

———. "Domestic UDMH Production in the DPRK." *Arms Control Wonk*, September 27, 2017. www.armscontrolwonk.com/archive/1204170/domestic-udmh-production-in-the-dprk/.

Melvin, Curtis, and Joseph S. Bermudez Jr. "KPA Navy Upgrades in the East Sea." *38 North*, September 1, 2016. www.38north.org/2016/09/munchon090116/.

Missile Defense Project. "Hwasong-5 ('Scud B' Variant)." *Missile Threat*, August 8, 2016. https://missilethreat.csis.org/missile/hwasong-5/.

———. "Hwasong-6 ('Scud C' Variant)." *Missile Threat*, August 8, 2016. https://missilethreat.csis.org/missile/hwasong-6/.

———. "Hwasong-9 (Scud-ER)." *Missile Threat*, August 8, 2016. https://missilethreat.csis.org/missile/scud-er/.

———. "Hwasong-12." *Missile Threat*, May 16, 2017. https://missilethreat.csis.org/missile/hwasong-12/.

———. "Hwasong-14 (KN-20)." *Missile Threat*, July 27, 2017. https://missilethreat.csis.org/missile/hwasong-14/.

———. "Hwasong-15 (KN-22)." *Missile Threat*, December 7, 2017. https://missilethreat.csis.org/missile/hwasong-15-kn-22/.

———. "KN-08/Hwasong 13." *Missile Threat*, August 8, 2016. https://missilethreat.csis.org/missile/kn-08/.
———. "KN-14 (KN-08 Mod 2)." *Missile Threat*, August 8, 2016. https://missile-threat.csis.org/missile/kn-14/.
———. "KN-18 (Scud MaRV Variant)," *Missile Threat*, April 18, 2017. https://missilethreat.csis.org/missile/kn-18-marv-scud-variant/.
———. "KN-23." *Missile Threat*, July 1, 2019. https://missilethreat.csis.org/missile/kn-23/.
———. "KN-24." *Missile Threat*, June 23, 2020. https://missilethreat.csis.org/missile/kn-24/.
———. "KN-25." *Missile Threat*, June 23, 2020. https://missilethreat.csis.org/missile/kn-25/.
———. "No Dong 1." *Missile Threat*, August 9, 2016. https://missilethreat.csis.org/missile/no-dong/.
———. "Pukguksong-2 (KN-15)." *Missile Threat*, March 5, 2017. https://missilethreat.csis.org/missile/pukkuksong-2/.
———. "Pukguksong-3 (KN-26)." *Missile Threat*, October 7, 2019. https://missilethreat.csis.org/missile/pukguksong-3/.
"N.Korea Steps Up Air Force Training Flights." *Chosun Ilbo*, May 29, 2012. http://english.chosun.com/site/data/html_dir/2012/03/29/2012032901309.html.
O'Carroll, Chad. "Exclusive: New Low-Visibility Corvette Spotted in North Korea." *NK News*, November 8, 2016. www.nknews.org/2016/11/exclusive-new-low-visibility-corvette-spotted-in-north-korea/.
Oliemans, Joost, and Stijn Mitzer. "A Navy Reborn: New Warships Spotted in North Korea." *NK Pro*, November 8, 2016. https://www.nknews.org/pro/a-navy-reborn-new-warships-spotted-in-north-korea/.
Pabian, Frank V., Joseph S. Bermudez Jr, and Jack Liu. "North Korea's Punggye-ri Nuclear Test Site: Satellite Imagery Shows Post-Test Effects and New Activity in Alternate Tunnel Portal Areas." *38 North*, September 12, 2017. 38north.org/2017/09/punggye091217/.
Panda, Ankit. "The Sinpo-C-Class: A New North Korean Ballistic Missile Submarine Is Under Construction." *The Diplomat*, October 18, 2017. https://thediplomat.com/2017/10/the-sinpo-c-class-a-new-north-korean-ballistic-missile-submarine-is-under-construction/.
Panda, Ankit, and Dave Schmerler. "When a North Korean Missile Accidentally Hit a North Korean City." *The Diplomat*, January 3, 2018. https://thediplomat.com/2018/01/when-a-north-korean-missile-accidentally-hit-a-north-korean-city/.
Posen, Barry R. "Nationalism, the Mass Army, and Military Power." *International Security* 18, no. 2 (1993): 80–124.
Raska, Michael. "North Korea's Evolving Cyber Strategies: Continuity and Change." *SIRIUS* 4, no. 2 (2020).
Rodong Sinmun, "uri sik sahoijueuigeonseoleul sae seungriero indohaneun widaehan tujaenggangryeong joseon rodongdang je8chadaehoieseo hasin gyeongaehaneun Kim Jong Un dongjieui bogoe daehayeo [Great Programme for Struggle Leading Korean-style Socialist Construction to Fresh Victory On Report Made by Supreme Leader Kim Jong Un at Eighth Congress of WPK]," *Rodong Sinmun*, January 9, 2021.
ROK Ministry of National Defense. *Defense White Paper 2018*. Seoul, ROK: ROK Ministry of National Defense, 2018.

ROK Ministry of Unification. *Understanding North Korea*. Seoul, ROK: ROK Ministry of Unification, 2017.

Scalapino, Robert A., and Chong-sik Lee. *Communism in Korea*. Berkeley, CA: University of California Press, 1972.

Scobell, Andrew, and John M. Sanford. *North Korea's Military Threat: Pyongyang's Conventional Forces, Weapons of Mass Destruction, and Ballistic Missiles*. Carlisle, PA: US Army War College Strategic Studies Institute, 2007.

Sutton, H. I. "North Korean VSV." *Covert Shores*, September 28, 2016. www.hisutton.com/North_Korean_VSV.html.

Tsukamoto, Katsuichi. *kitachousengunto seiji [The North Korean Army and Politics]*. Tokyo, Japan: Hara Shobo, 2000.

US Department of the Army. *ATP 7–100.2: North Korean Tactics*. Washington, DC: US Department of the Army, July 2020.

US Senate Committee on Governmental Affairs Subcommittee on International Security, Proliferation and Federal Services. *Senate Hearing 105–241: North Korean Missile Proliferation*, October 21, 1997.

Yoo, Yong-won. "bukhan bidaeching jeonryeok [North Korea's Assymetric Military Capabilities]." In *bukhangun sikeurit ripoteu [North Korea Military Secret Report]*, edited by Yong-won Yoo, Beom-chul Shin and Jin-a Kim. Seoul, ROK: Planet Media, 2013a.

———. "bukhan jaeraesik gunsaryeok mit mugichegye [North Korea's Conventional Military Force and Weapon Systems]." In *bukhangun sikeurit ripoteu [North Korea Military Secret Report]*, edited by Yong-won Yoo, Beom-chul Shin, and Jin-a Kim. Seoul, ROK: Planet Media, 2013b.

7 Understanding the threat

At the military parade held at Kim Il Sung Square on 15 April 2012, Kim Jong-un gave his first public speech, where he talked about the achievements of Kim Il-sung and Kim Jong-il in building the armed forces and vowed to further modernize and sharpen its readiness.[1] Aside from the ideological and propagandistic bravado, Kim Jong-un's speech reflected his faith in the developments in the KPA under the auspices of the Line of Self-Reliant Defence and his confidence in taking the next steps. Indeed, the problems in the KPA's readiness render the image that they are incapable of achieving the DPRK's strategic aim of unifying the Korean peninsula through force, let alone winning a war against the US and its allies. Still, that does not mean that the KPA is harmless. The nature of the DPRK and the KPA combined with the trajectory of the recent developments raise concerns not only in terms of the military threat, but also weapons proliferation, and also the future of the Stalinist state.

Military threat

To understand the nature of the North Korean military threat, one must be versed in the operational art that are designed to maximize the KPA's effectiveness while also compensating its readiness shortfalls. The DPRK's "military strategy" is based on three pillars: "surprise attacks," "quick and decisive wars," and "mixed tactics."[2] Based on the military strategy, the KPA's "principles of war" consist of "two-front war," "surprise," "mass and dispersion," "maneuverability," "initiative," "operational security," "annihilation," "combined operations," "mobility," and "rear area protection."[3] Finally, the "tactical doctrines" include: "sustainment;" "camouflage, concealment, cover, and deception;" "echelon forces;" "KPAAF and KPAN employment;" and "terrain appreciation."[4] Taken together, the military strategies, warfighting principles, and tactical doctrines create a formula that satisfies Friedman's criteria of effective tactics, being the combination of mental (deception, surprise, confusion, and shock) and physical (maneuver, mass, firepower, and tempo) tenets.[5] For the DPRK, designing its operational art according to the aforementioned criteria have been and will continue to be critical in confronting technologically superior opponents.

Based on the KPA's readiness and its warfighting concepts and doctrines, geographical nature of the Korean peninsula, as well as the location and readiness of the US, ROK, and Japanese forces, it is possible to broadly calculate and sketch out the KPA's battle drills, types of actions, execution procedures should conflict take place. The DPRK is set to conduct surprise attacks with fast maneuvers and tempo to execute their "two front war." At the front, the KPA will conduct massed swarm attacks combined with direct and indirect fire to penetrate the US and ROK's defense lines and destroy high-value, high-payoff targets including C4ISTAR and logistics systems, observation posts, key military units, areas of defense, and enemy reinforcements. Then in the rear, special operations and naval assets will be deployed deep into South Korea to capture, disrupt, and sabotage government (including military and law enforcement) assets and critical infrastructures (e.g. energy, financial, medical, transport and logistics, water, etc.), as well as assassinations, abductions, reconnaissance, and other counterstability operations. As Hodge opines, "the KPA has had decades to develop a campaign plan with a small number of military objectives that is probably extensively scripted and wargamed and would require limited flexibility and modification."[6] Still, despite the high degree of consistency in the fundamental aspects of the DPRK's operational art, the functional tactics can be executed with some variety depending on the purpose and nature of the mission, state of the opponent, state of its own forces as well as circumstances and environments such as location and time.

It is also important to understand how the DPRK has demonstrated a great deal of flexibility and adaptability. The KPA's tactics and war plans have changed over time, being a culmination of lessons and observations of various conflicts and operations. On one level, there are the firsthand lessons from the wars and operations by the North Koreans themselves, including Kim Il-sung's anti-Japanese partisan campaign, the Korean War, and the various armed attacks and provocations by the KPA since the armistice. Then on another level, there are observations of conflicts and military operations abroad, including but not limited to the two world wars, Chinese Civil War, Vietnam War, Arab-Israeli War, Kosovo War, Operation Desert Storm, Operation Iraqi Freedom, Syrian Civil War, operations by the Islamic State of Iraq and the Levant, Russo-Ukrainian conflict, and many others. Two underlying study themes become apparent. One is the strategies and tactics – particularly hybrid warfare – to effectively fight against technologically superior opponents and/or under disadvantageous circumstances. The other is the readiness and actions of the US, ROK, and Japan to identify not only their strengths and weaknesses but also the characteristics of their actions.

But the most vital aspect is that the DPRK continues to reconfigure its war plans that take into account of both the shortfalls and developments in the KPA's readiness. First, despite the outstanding readiness challenges, the DPRK has indeed modernized and diversified the KPA inventory. On top of some upgrades in conventional capabilities, the DPRK has acquired capabilities for cyber and electronic warfare, UAS, and strategic weapons of various types. While far from complete, the new capabilities allow the KPA to develop and execute its strategies, warfighting principles, and tactical doctrines in new and more diversified ways. For instance, ballistic missiles allow the DPRK to strike US and Japanese forces in the western Pacific, but also the US mainland. As for UAS, they could be utilized not only for swarm attacks against enemy units but also to spread chemical agents. Counterstability has also been enhanced with advancements in ICT, with the use of social media and other means to spread misleading information and subliminal messages to not only agitate pro-North Korean elements to join their cause but also create miscalculation, miscommunication, misinformation, misinterpretation, and misjudgment that would undermine response measures against the DPRK.

Second, the KPA still maintains its forward-deployed posture, with 70% of ground, 60% of naval, and 40% of air assets deployed below the Pyongyang–Wonsan line.[7] Moreover, many of the forward-deployed units are in HARTS for not only surprise attacks but also concealment and protection from enemy strikes. The forward-deployed nature allows the units to have a dual function. Offensively, the forward-deployed units are set up for surprise *blitzkrieg*-type assaults and delivery of intense fire to penetrate and force their way through the ROK and US defenses while also capturing major infrastructures in the Seoul Metropolitan area. Defensively, the forward-deployed units act as a tripwire against the ROK and US offensive maneuvers.

Third, there has been some level of configuration of air and naval capabilities for anti-access and area-denial. While the technological level of the KPA's air and naval capabilities makes them far from capable of neutralizing the US, ROK, and Japanese counterparts, they are sufficient in causing major disruptions particularly in areas close to the Korean peninsula. Indeed, such arrangements are not new, given the lessons the DPRK learned during the Korean War when they suffered major setbacks from the amphibious landings, and air and naval supremacy by the UN forces. Yet today, the KPA has a range of anti-air and anti-ship systems that are able to disrupt or at least harass incoming forces. Even for area-denial, while the DPRK quantitatively lacks long-range weaponry, the development of the Kumsong-3 ASCM indicates Pyongyang's efforts to engage naval forces approaching the Korean peninsula. Moreover, given the nature of the coastlines of the Korean peninsula and adjacent seas, the KPAN is capable of conducting operations such as minelaying as well as blockage and disruption of ports and sea lanes. Such strategies and tactics are also applicable for offensive operations such as providing cover for ballistic missile launches and also armed raids against the South – particularly on the western coast where KPA assets are extremely close to South Korean territory.

Fourth, notable developments are seen in jointness to enhance the KPA's capabilities for cross-domain warfare. The diversification of capabilities in recent years inevitably required the DPRK to coordinate the various assets for effective operations. The DPRK's renewed emphasis on cross-domain warfare was revealed in January 2010 when it conducted the Combined Maneuvers exercise that involved the ground, naval, and air units of the KPA. The Combined Maneuvers was the first publicized major joint exercise since Kim Jong-il became the SCAF and NDC Chairman, and, since 2010, the KPA is known to have conducted a number of such exercises. Although the exact details of the DPRK's joint operations remain unknown, the developments in KPA's readiness suggest that efforts have been made to improve the level of jointness and probably have conducted command post exercises that integrate the new capabilities – cyber, electronic, and WMD – into the military's operations.

Fifth, the asymmetric balance of forces have allowed the KPA to inflict cost-imposing measures against technologically superior opponents, creating what Scobell and Sanford correctly describes as a "logistical ordnance nightmare."[8] While the US, ROK, and Japan have benefitted much from next-generation technologies, many of the new assets are expensive, increasing the "cost-per-unit" or "cost-per-shot" ratio. Consequently, the US, ROK, and Japan are forced to use expensive armaments against the mass quantity of KPA platforms that are technologically inferior but cheaper to acquire and operate.[9] In offensive operations against the DPRK, Pyongyang's cost-imposing measures such as the various underground infrastructures create ISTAR challenges in detecting, locating, tracking, and engaging the KPA's assets. As for defense, the US, ROK, and Japan will be forced to use advanced air-to-air and surface-to-air missiles against a swarm of dated KPAAF aircraft that can be used as decoys or as intentional targets to "soak up vital inventories of interceptor missiles."[10] For instance, the KPA may deploy dozens of A-2 aircraft with the expectation that a good number of those will be shot down but counts on the surviving few to execute the attack. Indeed, the cost-imposing measures do not completely alleviate the capability deficiencies of the KPA, yet they are nonetheless vital means for the DPRK to create asymmetric effects against technologically superior opponents. Thus, even though the KPA's weapons systems are inferior in firepower, precision, and range, they are still capable of inflicting considerable damage if executed under the right conditions and tactics.

The advancements in, and diversification of capabilities – namely cyber and electronic warfare capabilities, and WMD – have evolutionized and further contextualized the KPA's warfighting concepts and doctrines. For example, in the event of a war, the KPA will attempt to break open the defenses of the ROK and US Forces Korea with artillery, MLRS, and SRBM while cyber and electronic attacks will be executed to disrupt or paralyze the US and ROK networks. Commandos will be inserted to the rear areas for assassinations, abductions, reconnaissance, sabotage, and other counterstability operations. At the same time, the ASCM, ICBM, IRBM, and SLBM will be used not only against the US and Japan for strategic strikes but also to attack the US Navy Seventh Fleet and the US Marine Corp before they reach Korean shores. Even if the DPRK does not execute the aforementioned coordinated attacks, the threat of such actions are used as deterrence measures against both preemptive and preventative attacks.

The developments in the DPRK's warfighting strategies and tactics, however, fall short of being magic wands to fix the KPA's weaknesses. As Friedman correctly argues, "irregular tactics are just tactics with a preference for maneuver, tempo, deception and surprise . . . to compensate for a lack of firepower, mass, and shock."[11] Even though the KPA has undergone some notable developments, readiness shortfalls still make them only effective under specific conditions. First, the focus on surprise attack and *blitzkrieg* for "decisive victory" means that the KPA needs to neutralize the threats in one shot while any imperfections could lead to severe countermeasures. Generally, states that rely on surprise attacks and fast wars are those that are either confident in their capacity to swiftly pulverize the enemy or those who are compelled to do so due to their limited logistical capacity to fight long conflicts. Although the DPRK has war plans and tactics designed for large-scale lethal strikes against the ROK and US forces, the shortfalls in force structural and operational readiness raise questions over their probability of success in winning the conflict. Moreover, it goes without saying that surprise attacks are most effective when the enemy least expects them. Yet in the case of the DPRK, the US, ROK, and Japan have grown familiar with the DPRK's patterns of behavior, leading to the strengthening of their readiness against possible abrupt actions.

Second, despite the DPRK's efforts to establish some kind of anti-access and area-denial arrangement, there are nonetheless major weaknesses in fighting beyond the confines of the Korean peninsula. Even if the DPRK's strategies are set on the Korean peninsula, the limited coverage still creates major vulnerabilities. Specifically, the KPAAF and KPAN's inability to secure air and maritime superiority creates major vulnerabilities. The shortfalls in firepower, power projection, and fuel inevitably limit their combat radius and ability to effectively deter and repel threats in advance. Moreover, the geographic dynamics set by the division has worked to the KPAN's disadvantage, where the two fleets are isolated from one another on the eastern and western coasts, making it impossible to group or come to one another's support without facing the US, ROK, and Japanese forces. In contrast, the air and naval units of the US, ROK, and Japan are armed with weaponry that boast greater lethality, precision, and range, and are also equipped with C4ISTAR systems that improve the effectiveness and efficiency of their operations. In addition, the US also has assets stationed in various parts of the globe that provides depth, enabling its ability to conduct a variety of strikes against the DPRK. Indeed, the DPRK is aware of the deficit, but even though Pyongyang has made some advancements in anti-air and anti-ship systems, as well as littoral combatants and submarines, the measures are still far short of filling the gaps and vulnerabilities.

Third, although the KPA may be able to work with scripted scenarios, there are questions over the ability to work beyond their playbook. Specifically, the severity of the KPA's readiness problems would be most apparent in full-scale conflict. As the US Chairman of the Joint Chiefs of Staff General Mark A. Milley said, "Wars are funny things. They have a logic all of their own and they rarely conform to preplanned timelines."[12] Thus, for the DPRK, there are questions over how they can handle long conflicts, particularly when the number of unexpected scenarios inevitably increase. The DPRK may switch to low-intensity warfare, but doing so would undermine much of the KPA's advantages and thus would merely prolong its eventual miserable fate as opposed to changing it.

Even if the KPA conducts southward attacks through *blitzkrieg* with heavy fire, the initial attacks to weaken the ROK and US defensive lines to cross the MDL will prove to be extremely resource-intensive, and the challenges will grow greater as the fighting progresses. The KPA would be compelled to call on the rear units or reserves, but the effectiveness of such measures are inevitably limited given that those units have been marginalized in resources and training. More importantly, the logistical demands would go well beyond the DPRK's capacity, with the poor quality and quantity of supplies, as well as questionable logistical supply chains that would exhaust the KPA. Problems would also be encountered in maintenance and repairs, where the KPA's force structural shortfalls indicate the high probability that a significant portion of platforms would be destroyed, critically damaged, or encounter mechanical problems. Thus, the question is how the KPA can attend the affected units and compensate the losses amid the chronic shortage of supplies and spare parts.

Indeed, the KPA could very well attempt to seize the resources and supplies in South Korea to feed and facilitate the KPA's operations.[13] But even if the KPA manages to secure such supplies, there are questions over whether they can establish effective and efficient logistical supply chains from the bases, as well as between and within the battlefields and units. In the extreme case, the KPA could be forced to take desperate measures such as conducting suicidal operations and making substandard weaponry from scrap metal and ancillary material.

The frictions and fog of war also apply to the state of personnel. Indeed, the KPA personnel are trained and indoctrinated to work under adverse conditions, and they are certainly tough in the way that they have lived through the dire circumstances in the DPRK amid the various widespread privations. But the magic of ideological discipline would ebb as the realities of war kicks in. The biggest psychological jolts would come from the firsthand encounters that contradict the propagandas of the regime, such as the fragility of the KPA and the overwhelming capability gap with the US, ROK, and Japan. Consequently, such problems would not only corrode the KPA's ability to fight in long-term conflicts but could also threaten the KPA's integrity and lead to misconduct by personnel.

The question is what the DPRK will do to overcome its readiness shortfalls. Naturally, the DPRK will be pressed to fill the KPA's readiness gaps, particularly as the US, ROK, and Japan have sharpened and strengthened their readiness against the DPRK. Certainly, it is easy to talk about the capabilities the KPA lacks – air and naval capabilities, missile defense systems, C4ISTAR systems, replenishment platforms, and many others. Such assets would provide the KPA with greater flexibility and effectiveness to conduct its operations. Yet it is critical to make the distinction between the capabilities the DPRK does not have, with the capabilities they need, and the capabilities they can get. The problem is not simply about economic capacity, but the fact that modernization aimed at matching technologically superior opponents could very well level the playing field in ways that undermine the KPA's asymmetric edge. For instance, if we take the extreme case where the DPRK pursues the acquisition of aircraft carriers, their capacity to acquire the vessel and form an adequate escort fleet are questionable, and even if they do so, whether they can or will be effectively used against the much superior air and naval capabilities of the US, ROK, and Japan are dubious at best.

Thus, one must not think too pervasively and argue that the DPRK can strengthen the KPA in any way they wish. Rather, discussions on how the DPRK will strengthen the KPA's readiness must consider the nature of not only Pyongyang's strategic goals and operational art but also its defense planning patterns. The recent developments in the KPA inventory and the military parade at the 75th anniversary of the WPK in October 2020 rendered the general direction of the DPRK's defense planning for the foreseeable future – strategic weapons (e.g. improved ballistic missiles with MaRV/MIRV capabilities), ground warfare, and capabilities for anti-access and area-denial. Even for naval platforms, on top of SSB/SSBN and some surface vessels, the DPRK would eye developments in manned and unmanned vessels for operations in the littorals with emphasis on speed and stealth. Cyber and electronic warfare capabilities will also continue to be pursued as means of disrupting and penetrating the networks of the US, ROK, and Japan. Developments in aerial capabilities are also possible although much would be needed before they can field any credible capabilities for air superiority. Other systems are also on Kim Jong-un's wish-list, including hypersonic weapons and advanced C4ISTAR systems, and that list is likely to grow in the coming years as the DPRK makes technological advancements. For example, the acquisition of long-range SAM such as the KN-06 indicates that the DPRK could develop its own ballistic missile defense system, although such prospects would pivot on whether Pyongyang can acquire and operate more advanced early warning systems.

The DPRK would also seek to gain and utilize new and emerging technologies that would enhance the KPA's readiness. On the high end, Artificial Intelligence (AI) and quantum computing would be of great interest for the DPRK to enhance the KPA's C4ISTAR systems as well as unmanned vehicles. Although both AI and quantum computing are both new and emerging technologies that even the most advanced militaries are still working on, the barriers in accessing and operationalizing those technologies even for the DPRK are not completely insurmountable. In addition, 3D printing would also be beneficial for logistics such as in producing spare parts within a short space of time. Indeed, the new and emerging technologies are still in their early stages, and there are nonetheless questions over how the DPRK will access, acquire, and adapt to these technologies. Nonetheless, the acquisition of the aforementioned technologies is much more likely and cost-effective than acquiring some of the high-end systems such as next-generation fighter jets, blue-water naval capabilities, and so forth.

Further developments in the KPA's capabilities will take place as long as the DPRK's strategic aims and objectives remain unchanged, and sticks to the belief that the military is the only effective tool it has. Even if the KPA does not undergo transformative changes with cutting-edge platforms, they are likely to follow the current trajectory of military modernization that adheres to the strategic, operational, and tactical doctrines designed to penetrate the vulnerabilities of technologically superior adversaries.

Weapons proliferation

The modernization of weaponry and need for hard currency to develop (or sustain) the economy raises concerns from the proliferation viewpoint. The modernizing state of North Korean weaponry ups the possibility of weapons proliferation. For the DPRK, there is greater marketability with the variety of updated systems, including not only conventional and strategic weapons but also C4ISTAR equipment. There would also be interest for buyers – both state and non-state actors – as the DPRK would offer an assortment of weapons that may not be the best but more accessible compared to other states.

Indeed, there are a number of factors that inhibit Pyongyang's arms trade, namely the assortment of bilateral and multilateral sanctions, stricter trade control regimes, and greater surveillance of movements in and out of the DPRK. Still, the aforementioned measures have been far from perfect, evidenced by the variety of tactics the DPRK has employed to dodge sanctions, including shipments via third-party contractors, front companies of the MND and WPK, and ship-to-ship transfers. Moreover, arms trade is also not only limited to physical shipments, as the DPRK could also share knowledge through technical assistance via technicians, and also by selling weapons designs online. While we have gained greater knowledge about Pyongyang's illicit transactions, the actual network and activities are likely to be much more diverse, expansive, and sophisticated.

The implications on international security are obvious, where North Korean weapons could reach state and non-state actors in various parts of the globe but with strong intent on using them. Even in the North Korean context there are grave concerns should its arms sales prove to be successful, where the revenue could be used to develop the state economy, but more likely for the military sector to finance further R&D and production. While it is premature to say that the DPRK will become a major weapons producer and supplier, the developments certainly warrant greater attention.

The fate of North Korea and the KPA

The magnitude and nature of the "North Korean threat" and the future of the KPA are much dependent on the fate of the regime. Of course, predicting the end of the regime is impossible, and one can only settle with the hypothesis that the regime will survive as long as the three principles of centralization, politicization, and inheritance remain intact. Yet still, while the centralized and politicized command and control over defense planning and the KPA have helped the Kim dynasty's survival, the structure and processes would reveal its dire side effects should the regime's authority weaken.

Various scholars have talked about the possible scenarios of regime collapse in North Korea.[14] While discussions concerning the possible end of the regime are important, it is also critical to think about why it has not collapsed. As Cha correctly notes, "North Korea has survived as the impossible state because no one on the inside is empowered to overthrow it, and no one on the outside cares enough to risk the costs of changing it."[15] Even regarding assassinations and coup d'états, the North Korean regime has always been conscious about the risks of the military turning against them and have implemented various mechanisms to preemptively and preventatively quell such possibilities, epitomized by the multi-dimensional command and control over the armed forces, as well as the various internal security apparatuses such as the MSC.[16] Thus, realistically, a coup d'état is only possible if the above measures are significantly weakened.

Similar arguments also apply to the prospects of reform. As Lankov accurately described, the DPRK has the choice of either "chemotherapy" or "surgery," where current measures would prolong survival for some time but with negative long-term results, while bold measures for real recovery and growth would have lower chances of success.[17] For now, Kim Jong-un has focused on survival through modernization without reform, building a totalitarian form of authoritarian developmentalism. Like his predecessors, Kim Jong-un has averted reform out of the fear that it would weaken the regime's control while changing the populace's consciousness that leads to demands for further changes. Even for the officials, they fear being made responsible for any failures, particularly after witnessing the execution of officials who were supposedly in charge of the failed currency redenomination in 2009. For the citizens, even though many have lost faith in the old Stalinist system and have resorted to their own measures for survival, none of the rule-bending behaviors are politicized or have even remotely led to attempts to overthrow the regime as they are focusing on their short-term economic welfare.

Questions concerning Kim Jong-un's health have led to questions about what could happen should the leader become suddenly incapacitated. In any event, the future of the DPRK would much depend on how the WPK and the KPA react to fill the vacuum, and whether the succeeding regime will inherit the policies of the old system or try to revise it. If the WPK's rule remains in place, then we could see either a system that is run by someone from the Kim family or some kind of collective, one-party dictatorship like China. Yet whatever the case, the bigger question is whether the succeeding regime would be able to effectively operate the political system and exercise adequate command and control over the armed forces.

The major question is how the KPA will view and respond to the political developments. The new leadership will know very well that any policies that could be perceived as neglecting or making the military more vulnerable would lead to serious repercussions. Despite Pyongyang's efforts, the serious decline in the military personnel's welfare and increase in misconduct is proof that Songun has failed to provide adequate material support to the officers and soldiers. Even if the military has not turned against the Kim dynasty, much could change depending on the developments and uncertainties that unfold in the post-Kim Jong-un regime. Should at any point the KPA no longer view the new regime as credible or trustworthy enough to consign their future, the military could attempt to take control. Naturally, the new leadership would do what it can to ensure the military's continued loyalty but that would require the KPA to be regarded as *the* institution for national security as well as being treated as a prioritized institution – meaning a return to Songun, but on a higher level.

Greater problems will erupt should there be any serious forms of political instability. The centralized system combined with the check-balance mechanisms among the various components of the armed forces with limited level of horizontal integration indicates that one slight splinter or imbalance in the command and control structure would fragment the very system Kim Il-sung and his successors crafted. Under this scenario, the military could fragment according to their positions and status, but also divergent interests and ends, including, but not limited to: regime loyalists who attempt to restore the old system; reformists who seek to establish a new type of government focusing on growth and perhaps even democratic values; and hardliners who view that the old regime was too soft, therefore, seeking to establish a militarist regime. Moreover, there could be groups of vigilantes that seek to restore order in their own locales as well as bandits that seek to exploit the chaotic circumstances. The situation could see the DPRK slip into a warlord era with the risks of a bilateral or worse, multilateral civil war, with great uncertainties over who will take hold of the deadly strategic arsenals that once belonged to the KPA.

Of course, one could talk about the possibility of democratization in North Korea. However, the problem is that the North Korean society has never experienced or has knowledge about the mechanisms of democratic reform and governance. Even if a growing number of citizens have learned that life is better in democratic states, they have no knowledge about how those were attained. Indeed, one could talk about North Korea democratizing in the same way as South Korea. Yet the road to democracy in South Korea took place over the course of decades that included many trials and errors, ousting of leaders, coup d'états, and, most importantly, the constant presence of pro-democracy groups and activists. The political conditions in North Korea share none of those features. Thus, although democratization is certainly not impossible, the process will be long, and as long as there are no clear visions and conditions for democratization (or at least better governance) the political uncertainties in North Korea would only grow.

Regional implications and the need for new strategies

Although the findings of this study suggest that the KPA's readiness is far from sufficient in winning a war and/or unifying the Korean peninsula through force, the KPA is still capable of inflicting considerable damage, and Pyongyang has not dialed back in using the threat of force. Rather, the DPRK has and will continue to strengthen the KPA's readiness while also strengthening the regime's grip over the armed forces, presenting numerous implications in the regional security context. Even if the DPRK focuses on more defensive ends for regime survival, they will continue to pursue those ends offensively, using force as a means of molding favorable conditions. Kim Jong-un's speech and the amendments to the WPK memorandum at the Eighth WPK Congress in January 2021 revealed once again that the DPRK prioritizes and has strong faith in its military to achieve the state's strategic ends. Thus, as long as the DPRK remains to be who they are, there will be no change to the existential threat they pose.

The combination of the threat posed by the KPA and uncertainties over the regime's fate require new strategies to deal with the DPRK. Proposing the specific strategies and tactics to effectively deal with the North Korean threat is beyond the scope of this study and is reserved for further discussions among strategists, policymakers, scholars, and thinktankers. However, it must be noted that there is no one-size-fits-all strategy to deal with the DPRK. Thus, as a starting point, discussions are needed on the key dilemmas and problems in the measures that have been proposed or practiced to date.

One key problem is that all too often, the international community and regional stakeholders have been overcaptivated by the symptoms, rather than the causes of the threat posed by the DPRK and the conflict on the Korean peninsula. While there is no denying that the division of the Korean peninsula began from geopolitical power plays between the US and the USSR, the Korean War and the continued division and conflict as well as the human rights abuses are blamed on the DPRK regime that has focused on its own survival rather than democracy and peace. Such problems raise questions concerning the means and ends to peace. Although there is no doubt that a treaty or agreement to end the Korean War is a crucial step toward peace and eventual unification, the actual outcome is much dependent on the terms and conditions. On the one hand, some may argue that unconditionally ending the Korean War will be a key step toward peace and solving the military threat. But the problem is that doing so would not deal with the causes of the conflict and other vital issues such as human rights, but also Pyongyang could very well shred the agreement if it ever feels it is not making any gains. On the other hand, the more thorough approach would be one that makes the DPRK: irreversibly disarm its WMD and dismantle its production capacity; retracts its forward-deployed forces; takes responsibility for the Korean War and the myriad breaches of the armistice since 1953; and addresses its human rights abuses. The problem, of course, is that Pyongyang will simply refuse and most likely respond in a bellicose manner.

Diplomatic solutions are vital, and there have been various bilateral and multilateral efforts to kick-start dialogues and negotiations with Pyongyang albeit many of them have failed to go beyond a certain point in bringing substantive and sustainable solutions. Indeed, both Moon Jae-in and Donald Trump were able to set up summits with Kim Jong-un in 2018 that presented themselves as prime opportunities for diplomatic breakthroughs.[18] However, the problem has never been about whether dialogues and summits can be held or not, but whether they can be held under the right conditions and pursued with the right strategies that lead to meaningful outcomes. The DPRK may start negotiations based on vague principles but will bring clear demands for normalization of relations with the US, as well as security guarantees including the lifting of sanctions and scaling back defensive measures against Pyongyang. Although some may consider such demands as affordable steps for further progress, the issue is how and whether the DPRK will take credible reciprocal measures, particularly considering Pyongyang's notoriety for dishonoring agreements and reverting to bellicose behavior. The problem here is not simply about the DPRK's ability to cheat deals but also the conflicting interpretations and standpoints over the sequencing and verification of measures proposed and negotiated.[19]

Sanctions too have been questionable in their effect. While the collection of bilateral and multilateral sanctions has certainly hurt the North Korean economy, they have failed in changing Pyongyang's behavior, let alone slowing its military modernization program. Questions over the sanctions vis-à-vis the DPRK are four-fold: first, the DPRK over the years has built various measures to ensure the regime's survivability against sanctions; second, sanctions have led to sharper reactions from Pyongyang and have also been used to shore up legitimacy by blaming the economic suffering on the states imposing the sanctions; third, the DPRK has creatively evaded enforcement measures, including not only the employment of front companies and third parties but also ship-to-ship transfers; and fourth, the effectiveness of sanctions will also ebb as states would eventually run out of the items and options to sanction the DPRK. That said, lifting sanctions before any positive action from the DPRK will only set a bad precedent that blackmail works, providing opportunities for further exploitation.

Given the nature of the threat posed, military measures to deter – and if necessary, defend – against the DPRK are critical. Discussions on measures needed against the KPA's threat have often involved capabilities, including enhanced C4ISTAR, ASW, ballistic missile defense, and, finally, strategic and tactical strikes. However, the more important questions are not about the capabilities of the US, ROK, and Japanese forces but their application. Although one could argue that the deterrence strategies of the US and its allies have been effective in preventing full-scale attacks by the DPRK, the "gray-zone" situations caused by Pyongyang's military actions and their threats to use force demonstrate the shortfalls of the measures to date. To enhance coercion against the DPRK, the most important is to target Pyongyang's "center of gravity" to demonstrate the ability to inflict pain. Such measures would require greater emphasis on deterrence by punishment rather than denial, including the ability and intent to conduct preemptive and/or even preventative strikes should the DPRK dare to take any deadly action. Certainly, the shift toward more offensive strategies would trigger a tough response from the DPRK, leading to new security dilemmas. Nevertheless, the consequences of not enhancing the deterrence measures are greater than not doing anything at all, as the latter would only allow Pyongyang to continue to push the envelope and enhance their military leverage.

Discussions are also needed on intervention should the North Korean regime collapse or shows signs of major instability that affects regional security. Indeed, the US and ROK already have contingency plans such as the highly classified Conceptual Plan 5029 (CONPLAN 5029) covering six scenarios relating to major instabilities in the DPRK, including: coup d'état; civil war; mass revolt against the regime; rebel forces gaining control of WMD; massive exodus of refugees;

major natural disasters; and others. Yet CONPLAN 5029 is largely conceptual and extremely controversial for both political and practical reasons. In practice, there is a constellation of scenarios that could unfold should there be some kind of instability in the DPRK. The major problem here is not just the scenario that requires intervention, but the challenges that will surface during and after the intervention. Intervention itself does *not* establish democracy or unification but is merely a step toward setting the conditions that facilitate developments toward those goals. The stabilization and clean up process would be a mess, and considerable time would be needed to disarm what was formerly the KPA, and also in dealing with the political, economic, and societal fallout. In addition, there is no guarantee that intervention will be welcomed with open arms. Since the formative years, the DPRK regime has indoctrinated both the military and the society that the US and its allies are readying to pounce on North Korea. Thus, if the ROK and US forces unilaterally enter North Korea, the policies and rhetoric of the regime would be proven right, consequently triggering not only a defensive response by the KPA, but also possibly by insurgents.

Effectively dealing with the North Korean threat depends much on the US–ROK alliance. Although the alliance has stayed, ambiguities and strategic disconnects between the two have become increasingly apparent under the progressive administrations in the ROK. Greater problems are also seen in cooperation and coordination between the US, ROK, and Japan. The most pressing areas of US–ROK–Japan cooperation would be missile defense, ASW, and humanitarian assistance and disaster relief.[20] The problem, however, is that while the US, ROK, and Japan have expressed their intentions to cooperate, they have seldom been in lockstep in dealing with the DPRK due to increasingly divergent strategic agendas and perceptions. Problems are also compounded by the troubling state of Japan–ROK relations caused by not only historical animosities but also colliding national identity discourses.[21] Despite the problems, trilateral cooperation and coordination are vital to comprehensively and effectively deal with not only the North Korean threat but also regional stability. Moreover, the trilateral mechanism will be a critical part of an Indo–Pacific-wide security pact, expanding the currently developing "quad" that includes the US, Japan, India, and Australia. The pact could further expand, including the United Kingdom, likeminded Southeast Asian states, and also Taiwan. The consequences of failing to enhance cooperation are obvious, not only weakening the deterrence measures against the DPRK but also providing exploitable vulnerabilities to the DPRK, China, and Russia that would undermine regional security and stability.

There are major questions concerning China whose strategic interests collide with or at least is distant from those of the US and its allies. Above all, there is the geostrategic factor where Beijing aspires to become a regional hegemony. Even regarding the Korean peninsula, China prefers to keep the DPRK alive for two reasons, one to protect Chinese economic assets and interests in North Korea, and the other to maintain the strategic buffer zone between itself and the US–allied ROK. Indeed, China certainly does not encourage any unilateral aggression by the DPRK that would upset regional stability. Yet at the same time,

China would resist or cushion any hardline responses against the DPRK and would do what it can to ensure that its interests on the Korean peninsula remains uncompromised. Moreover, China would try to intervene faster than the ROK and US in the case of any major contingencies in North Korea, leading to a whole new host of problems on the Korean peninsula.

Formulating the right strategies to deal with the North Korean threat is far from easy. The overarching puzzle is how to deter and defend against the DPRK military threat while also averting any uncontrollable instabilities within the Stalinist state. Even in dealing with the military threat, neutralizing, or at least denying the KPA's capabilities are not hard, but neutralizing the DPRK's intentions will be difficult, as long as the regime maintains their character. The developments under Kim Jong-un have opened new chapters to the North Korea problem that are far from business as usual. We are now at a critical juncture that requires new strategies to deal with the threats while facilitating developments that will bring peace and save the North Korean citizens who have been enslaved by the Kim dynasty for over seven decades. When new, effective strategies are formulated and exercised against the DPRK, peace, stability and the long-awaited peaceful and democratic unification of the two Koreas will be achieved, leading to new steps in international peace and security.

Notes

1 Jong-un Kim, *songuneui gichireul deo nopi chukyeodeulgo choihuseungrireul hyanghayeo himchage ssawonagaja [Let Us March Forward Dynamically Towards Final Victory, Holding Higher the Banner of Songun] (15 April 2012)* (Pyongyang, DPRK: Joseon Rodongdang Chulpansa, 2013).
2 US Department of the Army, *ATP 7–100.2: North Korean Tactics* (Washington, DC: US Department of the Army, July 2020), 1.13.
3 Ibid., 1.14–1.16.
4 Ibid., 1.16–1.17.
5 B. A. Friedman, *On Tactics: A Theory of Victory in Battle* (Annapolis, MD: Naval Institute Press, 2017), 22. Friedman also argues that "the key to successful tactics is combining mass and firepower in a package that can operate at a fast tempo and thus outmaneuver enemy forces." Ibid., 206.
6 Homer T. Hodge, "North Korea's Military Strategy," *The US Army War College Quarterly: Parameters* 33, no. 1 (2003): 78.
7 Sung-bin Choi, Jae-moon Yoo, and Si-woo Kwak, *bukhan gunsusaneob gaehwang [Current State of the North Korean Military Industry]* (Seoul, ROK: Korea Institute for Defense Analyses, 2005), 12–14.
8 Andrew Scobell and John M. Sanford, *North Korea's Military Threat: Pyongyang's Conventional Forces, Weapons of Mass Destruction, and Ballistic Missiles* (Carlisle, PA: US Army War College Strategic Studies Institute, 2007), 54.
9 In reference to North Korean submarines, an engineer claimed that the US would not want to "risk a billion-dollar nuclear sub in the littoral." An engineer quoted in: P. W. Singer, *Wired for War: The Robotics Revolution and Conflict in the 21st Century* (New York, NY: Penguin Press, 2009), 226.
10 TNI Staff, "North Korea's Air Force is Total Junk (But it Can Still Kill)," *The National Interest*, October 7, 2017, http://nationalinterest.org/blog/the-buzz/north-koreas-air-force-total-junk-it-can-still-kill-22647.

11 Friedman, *On Tactics: A Theory of Victory in Battle*, 181.
12 Mark A. Milley, "Eisenhower Luncheon Keynote Address at the Association of the US Army," October 13, 2015, https://api.army.mil/e2/c/downloads/412559.pdf.
13 Scobell and Sanford, *North Korea's Military Threat: Pyongyang's Conventional Forces, Weapons of Mass Destruction, and Ballistic Missiles*, 65; US Department of the Army, *ATP 7–100.2: North Korean Tactics.*
14 For example, see: Bruce W. Bennett and Jennifer Lind, "The Collapse of North Korea: Military Missions and Requirements," *International Security* 36, no. 2 (2011).
15 Victor D. Cha, *The Impossible State: North Korea, Past and Future*, 1st ed. (New York, NY: Ecco, 2012), 12–13.
16 For a discussion on the possible scenarios of the assassination of Kim Jong-un, see: Sung-min Cho, "Anticipating and Preparing for the Potential Assassination of Kim Jong-Un," *International Journal of Korean Studies* 19, no. 1 (April 2015).
17 Andrei N. Lankov, *The Real North Korea: Life and Politics in the Failed Stalinist Utopia* (Oxford, UK: Oxford University Press, 2013), 165–66.
18 Regarding Trump, Kim Jong-un saw Trump's unconventional style in two ways, where on the one hand he may indeed order an attack against the DPRK, while on the other he could be lured into a deal that most US presidents would have refused.
19 Bill Dorman and Chris Vandercook, "Impact of DPRK – USA Summit (with Carl Baker, Charles Morrison, Dan Leaf)," *The Conversation*, June 13, 2018, hawaiipublicradio.org/post/conversation-impact-dprk-usa-summit.
20 Ryo Hinata-Yamaguchi, "Completing the US-Japan-Korea Alliance Triangle: Prospects and Issues in Japan-Korea Security Cooperation," *The Korean Journal of Defense Analysis* 28, no. 3 (Fall 2016).
21 Brad Glosserman and Scott Snyder, *The Japan-South Korea Identity Clash: East Asian Security and the United States* (New York, NY: Columbia University Press, 2015).

References

Bennett, Bruce W., and Jennifer Lind. "The Collapse of North Korea: Military Missions and Requirements." *International Security* 36, no. 2 (2011).

Cha, Victor D. *The Impossible State: North Korea, Past and Future.* 1st ed. New York, NY: Ecco, 2012.

Cho, Sung-min. "Anticipating and Preparing for the Potential Assassination of Kim Jong-Un." *International Journal of Korean Studies* 19, no. 1 (April 2015).

Choi, Sung-bin, Jae-moon Yoo, and Si-woo Kwak. *bukhan gunsusaneob gaehwang [Current State of the North Korean Military Industry].* Seoul, ROK: Korea Institute for Defense Analyses, 2005.

Dorman, Bill, and Chris Vandercook. "Impact of DPRK – USA Summit (with Carl Baker, Charles Morrison, Dan Leaf)." *The Conversation*, June 13, 2018. hawaiipublicradio.org/post/conversation-impact-dprk-usa-summit.

Friedman, B. A. *On Tactics: A Theory of Victory in Battle.* Annapolis, MD: Naval Institute Press, 2017.

Glosserman, Brad, and Scott Snyder. *The Japan-South Korea Identity Clash: East Asian Security and the United States.* New York, NY: Columbia University Press, 2015.

Hinata-Yamaguchi, Ryo. "Completing the US-Japan-Korea Alliance Triangle: Prospects and Issues in Japan-Korea Security Cooperation." *The Korean Journal of Defense Analysis* 28, no. 3 (Fall 2016).

Hodge, Homer T. "North Korea's Military Strategy." *The US Army War College Quarterly: Parameters* 33, no. 1 (2003).

Kim, Jong-un. *songuneui gichireul deo nopi chukyeodeulgo choihuseungrireul hyang-hayeo himchage ssawonagaja [Let Us March Forward Dynamically Towards Final Victory, Holding Higher the Banner of Songun] (15 April 2012)*. Pyongyang, DPRK: Joseon Rodongdang Chulpansa, 2013.

Lankov, Andrei N. *The Real North Korea: Life and Politics in the Failed Stalinist Utopia*. Oxford, UK: Oxford University Press, 2013.

Milley, Mark A. "Eisenhower Luncheon Keynote Address at the Association of the US Army." October 13, 2015. https://api.army.mil/e2/c/downloads/412559.pdf.

Scobell, Andrew, and John M. Sanford. *North Korea's Military Threat: Pyongyang's Conventional Forces, Weapons of Mass Destruction, and Ballistic Missiles*. Carlisle, PA: US Army War College Strategic Studies Institute, 2007.

Singer, P. W. *Wired for War: The Robotics Revolution and Conflict in the 21st Century*. New York, NY: Penguin Press, 2009.

TNI Staff. "North Korea's Air Force is Total Junk (But it Can Still Kill)." *The National Interest*, October 7, 2017. http://nationalinterest.org/blog/the-buzz/north-koreas-air-force-total-junk-it-can-still-kill-22647.

US Department of the Army. *ATP 7–100.2: North Korean Tactics*. Washington, DC: US Department of the Army, July 2020.

Index

Note: Page numbers in **bold** indicate a table on the corresponding page.